Caribbean Examinations Council®

Geography

CAPE®

UNIT 1

for self-study and distance learning

Originally published in 2010 by the Caribbean Examinations Council in association with The Commonwealth of Learning.

This material was originally developed by the Caribbean Examinations Council with assistance from the Commonwealth of Learning.

Repackaged and distributed in 2012 by:
Nelson Thornes Ltd
Delta Place
27 Bath Road
CHELTENHAM
GL53 7TH
United Kingdom

10 9 8 7 6 5 4 3 2

A catalogue record for this book is available from the British Library

ISBN 978 1 4085 1817 5

Cover Image by: Mark Lyndersay, Lyndersay Digital, Trinidad
www.lyndersaydigital.com

Page make-up by OKS Prepress, India

Illustrations by OKS Prepress, India and Dave Russell Illustration

Printed and bound by CPI Group (UK) Ltd, Croydon, CR0 4YY

The manufacturer's authorised representative in the EU for product safety is Oxford University Press España S.A. of El Parque Empresarial San Fernando de Henares, Avenida de Castilla, 2 – 28830 Madrid (www.oup.es/en or product.safety@oup.com). OUP España S.A. also acts as importer into Spain of products made by the manufacturer.

Contents

Acknowledgements

The author and the publisher would like to thank the following for permission to use or adapt copyright material:

Figure 3.1 taken from *Atlas for Caribbean Examinations*, (3rd Edition) published by Pearson Education (2007). Copyright © Collins Bartholomew Ltd. Reproduced with permission; Figure 5.1 Office for National Statistics © Crown Copyright 2011. Reproduced under PSI licence no. C2009002012; Tables 6.1, 6.2, 6.3, 6.4 from Growth of urban population by Development region' taken from *World Urbanization Prospects: The 2007 Revision*, United Nations Department of Economics and Social Affairs/Population Division © United Nations, New York, 2008. Reproduced with permission. Figure 13.1 AFP/Getty Images.

Every effort has been made to trace the copyright holders and we apologise if any have been overlooked. Should copyright have been unwittingly infringed in this book, the owners should contact the publishers who will make corrections at reprint.

Introduction

Purpose

The **Caribbean Examinations Council** (CXC®) has developed Self-Study Modules for a number of **Caribbean Secondary Education Certificate** (CSEC®) and **Caribbean Advanced Proficiency Examination** (CAPE®) subjects. The main purpose of the Modules is to provide both in-school and out-of-school candidates with resource materials which should help them in preparing for **CXC** examinations. Each Module is student centred and its language is student friendly.

The Geography Unit 1 course is designed for people over the age of sixteen who wish to further their studies. The course is equally useful to those who are pursuing part-time study and those enrolled in full-time education.

You may have completed five years of secondary education, or you may be a mature student with work experience. The course is based on the assumption that you are already able to do the following:

1 interpret, accurately, material intended for the informed lay-person;
2 clearly express personal opinions and factual information, demonstrating logical sequencing and appropriate English up to the level of the **CSEC** General Proficiency or its equivalent.

Course Aims

The course aims to enable students to:

1 develop an understanding of the location and distribution of geographic phenomena
2 develop an understanding of the nature of Physical and Human Geography and their interactions
3 explain the processes at work in Physical and Human Geography
4 develop an understanding of the environmental consequences of human action
5 develop an appreciation of the current social and economic problems in their geographical setting
6 encourage an appreciation of the dynamic nature of Geography
7 help in the understanding and application of spatial models and concepts to the study of Geography
8 develop an understanding of the range of techniques, the acquisition of practical skills, and an appreciation of information technology that enhance geographical knowledge
9 create awareness of the variety of Caribbean environments through field activities
10 promote knowledge and understanding of world geography
11 develop an understanding of the place of the Caribbean in the wider world
12 encourage a critical and reflective approach to the study of Geography.

Course Structure

The course consists of fourteen Modules, which are all based on the **CAPE** Geography Unit 1 Syllabus. Each Module addresses the skills and content of a specific Module of the Syllabus. However, the sequence of the Modules does not necessarily mirror that of the syllabus Modules since the syllabus Modules are not bound by a rigid sequence. The sequence of topics in this course is designed to facilitate study by leading you through topics in a way which will enable you to build on previously learnt skills.

What Resources Will You Need?

Remember that these Modules will not be all that you need to complete the syllabus and prepare for your examination. You are expected to make use of the resources listed at the end of the course book as well as engage in other wide, general, reading which will improve your general knowledge, vocabulary and structural competence.

You will also need basic study equipment, for example, paper, pens, pencils, and highlighters for marking important parts of the text. A good dictionary and a thesaurus are also essential to this programme.

Managing Your Time

Remember to put aside special time each day for general reading in addition to your study time.

Module Structure

Each Module is divided into seven sections to facilitate your study, as indicated below:

Introduction This places what you are about to study in the context of your everyday life and relates it to what you have done in previous Modules.

Content This lists the topics that are to be covered in the Module.

Objectives These help you to identify the specific skills that you should have acquired by the end of the Module. You should read these carefully to acquaint yourself with what you are meant to be learning during the Module.

Activities Instructions are provided at the start of each activity. Read all instructions carefully before you attempt the activity. Some activities require you to think about something before you read any further. You should take the necessary time to do so. The thinking activity is designed to help you focus your thoughts in the direction which will facilitate your ability to complete the activities that follow.

Feedback Each activity has a feedback section that allows you to determine how well you have done in the activity. If you have not completed the activity successfully, you should re-read the preceding examples or information carefully.

End Test This comes at the end of each Module and is designed to ensure that you have acquired those skills identified in the objectives. There is a feedback section following the End Test which allows you to measure the accuracy of your answers to the test so that you will know whether or not you have acquired the competencies. If there are questions in the End Test that you have not answered satisfactorily, ensure that you return to the relevant section of the Module and review those areas until you are satisfied that you have understood the concept.

Key Points These summarize important concepts that you need to remember and pay special attention to as you work through the course.

Examinations

You must ensure that you have access to the version of the syllabus that outlines the structure of the examination for the year in which you intend to sit it.

1 Population Distribution and Density

General objectives

On completion of this module you should:

- understand the factors affecting the distribution of human populations locally, regionally and internationally
- develop appropriate skills and techniques in Human Geography.

Specific objectives

You should be able to:

- explain the factors influencing population distribution using case studies
- assess the methods that depict population distribution.

Introduction

Population distribution refers to the way in which people are spread over an area. Some areas of the world are densely, and others sparsely, settled, that is, the population is unevenly distributed. Population distribution can be examined at different scales: globally, regionally and locally. The fundamental questions are where are they located and why there. It is important to understand the factors which influence population distribution. Both human and physical factors explain population distribution but, in general, physical factors have a greater influence in the less developed countries. The physical factors include climate, vegetation, relief, the availability of potable water, soils as well as diseases. Economic and political factors and communications are also important.

This module explains the concept of population distribution and distinguishes between the factors affecting such distributions at the global, regional and local levels. It focuses on the characteristics of population distributions that change in a spatial context, utilizing dot maps and the Lorenz curve. The module also outlines the merits and demerits of utilizing these methods of illustrating population distribution.

Content

- The factors influencing population distribution at the global scale.
- Case studies of factors affecting population distribution at the regional and local scales.
- Methods of depicting population distribution – dot map and Lorenz curve – and their merits and demerits.

Factors Influencing Population Distribution

Population distribution is the spread of people within an area, whether that area is a village, parish, county, country, continent or the globe. Populations are distributed unevenly across the globe. For example, according to Figure 1.1 on page 4, where each dot on the map represents 100,000 people, it can be recognised that populations are dispersed (widely scattered) in some areas and clustered (concentrated) in others.

In general, the map shows that most of the people live north of the equator which has a larger land area than the south, and within this area there is a marked concentration between 20° and 60° North. In addition, the margins of continents have attracted the greatest concentrations. Four clusters are apparent – East Asia (including China and Japan), South Asia (including India, Pakistan and Bangladesh) Europe and North Eastern United States of America and Canada. The Arctic North, the dry interior of Australia and the African Sahara have attracted relatively few people.

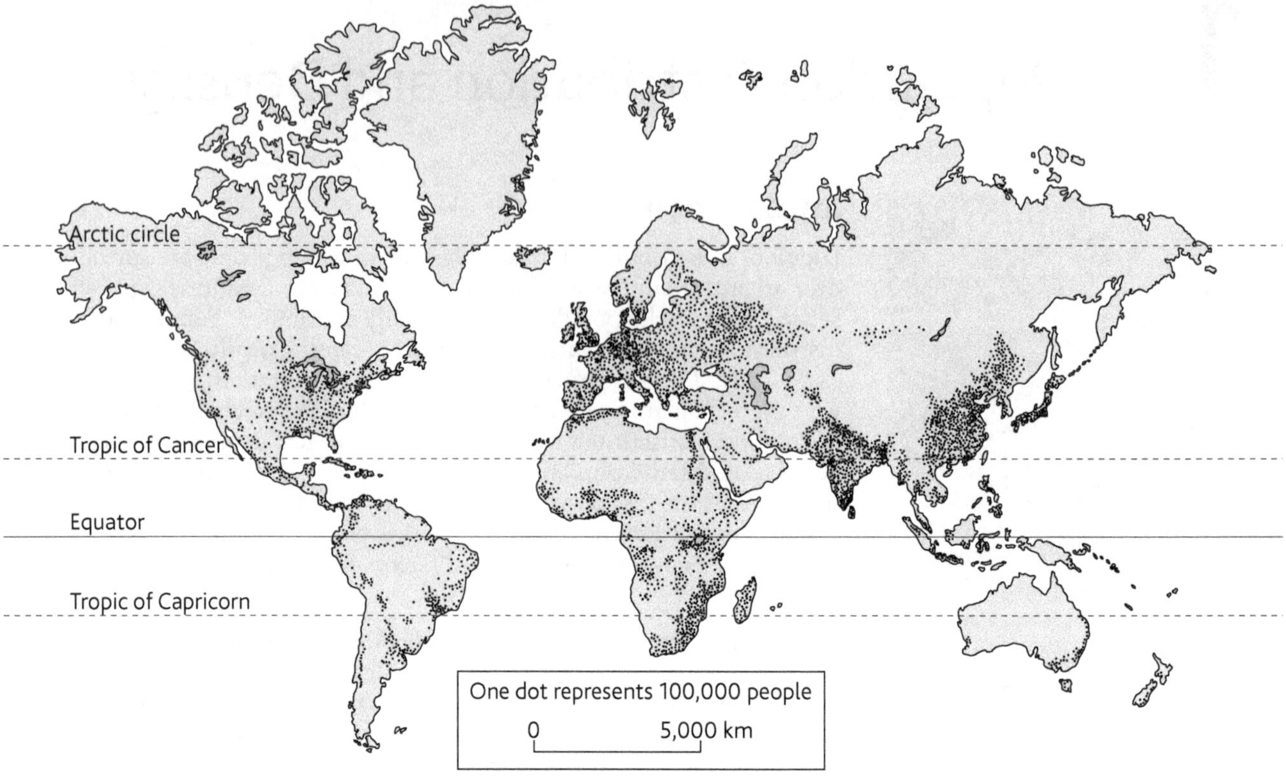

Figure 1.1 *Depicting World Population Distribution using a Dot Map*

Factors Affecting Population Distribution

Physical

Physical factors generally refer to natural or environmental influences. Modern technology can overcome many of the physical limits and has allowed settlement to spread over the earth.

- *Climate*: Climate affects the population distribution globally. Areas of extreme cold, such as the Arctic and Antarctic; and extreme aridity, such as the Sahara Desert, remain very sparsely populated. People tend to seek out climates that are not extreme. Therefore, temperate and tropical moist climates such as North West Europe and Asia tend to support the greatest clustering of population.
- *Soil*: Fertile soils for the cultivation of food have always attracted human settlement. River valleys in the past (the Nile and Tigris-Euphrates) as well as in the present have attracted large populations because of soils that are enriched by the rivers, access to potable water and food. The valleys of the large Asian rivers, such as the Ganges and Hwang Ho are some of the closely settled areas of the world.
- *Relief*: Low lying areas attract large populations. Highlands are less favoured because of their rugged nature, low temperatures, thin soils, steep slopes and short growing seasons.

- *Natural Resources*: some areas are either poor in natural resources or lack the capital to develop them. They therefore cannot support large populations. The large populations of Western Europe were sustained by the discovery of coal. The opening of the northern frontier in Mexico was given impetus by the silver deposits at Zacatecas. Mining settlements were also established in the Andes to exploit silver deposits. However, population numbers in mining areas tend to decline as deposits become expensive to mine or the mineral becomes depleted.

Human/Economic factors

- *Economic*: Economic considerations remain some of the most important factors in determining population distribution. Good infrastructure – roads, water, electricity – attract populations. Transportation costs make it difficult to exploit the resources found in the interior of South America. In general terms, the longer the journey into the interior, the greater the transportation costs and the less profitable economic operations in these parts are likely to be. Population distribution therefore remains low in the interior of South America.
- *Government policy*: There have also been cases in which governments have influenced population distribution patterns. The State may directly or indirectly decide to encourage development in a particular area or region. In modern times, governments have taken a regional view of resource development and created urban centres to act as economic nodes. For example, the Venezuelan government established the complex of towns that make up Cuidad Guayana on the Orinoco River to smelt bauxite and iron ore brought in from the Guiana Highlands to the south of the river. This development has also been influenced by the presence of a transportation route provided by the river. To encourage such developments they offer incentives, tax breaks and employment contracts to investors who are willing to go into these areas. Areas which receive high levels of investment are likely to attract large populations. Brasilia, the capital of Brazil, is an excellent example. The capital was transferred from Rio de Janeiro to a site that was nearer to the centre of the country to redistribute the population. It became the seat of government and headquarters of major Brazilian companies. Population growth was explosive. Planned for a half a million people, the city is now home to 2 million.
- *Communication*: Caribbean capitals developed because of their port functions, links between the colonies and Britain. Modern modes of transportation and communication have allowed population distribution to spread over most continents. Most places are easily accessible over land allowing individual choice in residence depending on personal preference. Some people are moving into previously sparsely settled areas because of highway construction, for example. New highways in Caribbean countries such as Highway 2000 in Jamaica and the Adams-Barrow-Cummins Highway in Barbados are changing the distribution of population.

Other factors which affect population distribution are historical factors, political instability, religious beliefs and cultural traditions.

Activity 1.1

Apart from the factors listed above, identify ONE other physical/environmental factor that may help to explain world population distribution.

Feedback

People tend to avoid heavily forested and marshy areas which often harbour diseases.

Case Studies of Factors Influencing Population Distribution at Regional and Local Levels Using Different Methods of Representing Population Distribution

Table 1.1 *Area and Population of Parishes of Barbados*

Parish	Population	Area (square km)
Christ Church	49,497	57
St Andrew	5,254	36
St George	17,868	44
St James	22,741	31
St John	8,873	34
St Joseph	6,805	26
St Lucy	9,328	36
St Michael	83,684	39
St Peter	10,699	34
St Philip	22,864	60
St Thomas	12,397	34

Source: National Census Report 2000, Barbados

Table 1.1 above shows data for the population of Barbados by parish. This data will be used to investigate the distribution of population in the island using a dot map and the Lorenz curve.

Dot maps

Dot maps usually display counts with each dot representing a specific value. They allow the reader to identify areas with higher or lower concentrations of people. In areas with a higher population frequency, dots are more numerous and appear closer together; sparser distribution is indicated when dots are more widely spread. It is visually very effective in identifying and locating variations in population distribution.

Steps in the construction of dot maps showing population distribution

1 Create a base map of Barbados showing the boundaries of the parishes in pencil.

2 Decide on the graphical size of the dot. Bear in mind the maximum and minimum number to be shown and the size of the map. A stencil could be used to ensure that the dots are of equal size.

3 Careful consideration must be given to the value of the dot. The largest population shown in the table is 83,684 and the lowest 5,254. A dot value of 2,000 may be chosen although it will distort the population distribution of the smaller parishes.

4 Calculate the number of dots needed for each parish and insert them in the appropriate parish on the map.

5 Consult topographical maps and other information to assist with the placing of the dots in the appropriate places in the parish.

The resultant dot map may look as shown in Figure 1.2.

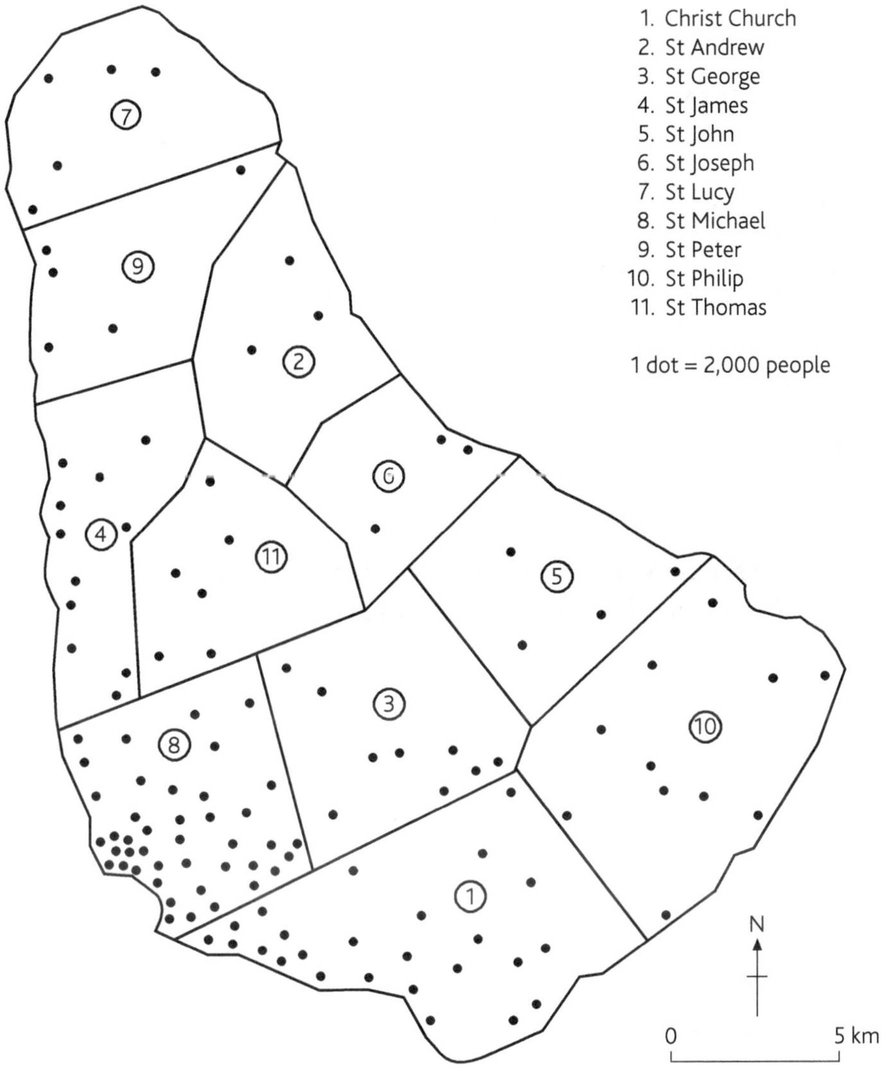

Figure 1.2 *Dot Map of Barbados showing population distribution*

Advantages of a Dot Map

- It is intuitive in associating the number of dots with number of people.
- It has a strong visual impact.
- It can recover original data from the map by counting dots.
- Manual placement of dots takes into account the distribution of other phenomena, such as agricultural land use, mountainous terrain.

Disadvantages of a Dot Map

- Dots may be interpreted as representing a single instance of the phenomena at a particular location.
- If dots form too dense a pattern – visual overlap – it is impossible to recover original data values.

Activity 1.2

1 Describe the population distribution shown on Figure 1.2.

2 Explain the factors influencing the population distribution of Barbados shown on Figure 1.2.

Feedback

1 The population of Barbados is very unevenly distributed. Most people are concentrated in the south west of the island, with very few in the east and north. The central area shows a moderate population spread.

2 The main factors accounting for the distribution of population in Barbados are the relief and socio-economic factors. The southwest of the island is the location of the capital, Bridgetown, and is the main port. It also has the greatest concentration of commercial and business activity which has attracted many persons seeking jobs. Along the south and west coasts intense tourism activity has attracted many people. To the east, the very rugged Scotland District is prone to landslides and has very few economic opportunities and, therefore, does not support many people. In the central area some land is still used for agriculture, although there are many recent housing developments.

- Human placement is subjective resulting in different distributions for same data.
- The poor choice of dot size can distort the accuracy of the map. Too small a dot value may create clustering, while too large a dot value may result in blank spaces appearing unpopulated.
- It is difficult to count large numbers of dots.

The Lorenz Curve

The Lorenz curve is a graph which represents inequality. It may be used to show inequality in the distribution of population, income, food, medical care and many other assets. When representing population distribution, population, shown on the x-axis, is plotted against area on the y. Both are plotted from 0 to 100 per cent. If the population were evenly distributed, then 20 per cent of the population would inhabit 20 per cent of the area; 75 per cent of the population would inhabit 75 per cent of the area. The relationship would be linear and the result would be a straight line. However, this rarely occurs as it has been shown that some areas are more attractive to settlement than others. The Lorenz curve commonly shows this unevenness. The greater the distance of the curve from the diagonal line, the greater the inequality in distribution.

Steps in the construction of Lorenz curve for Barbados

1 Calculate the percentage population and percentage area for each parish.

2 Order the parishes by decreasing size of population.

Parish	Population %	Cumulative %	Area %	Cumulative %
St Michael	33	33	9	9
Christ Church	20	53	13	22
St James	9	62	7	29
St Philip	9	71	14	43
St George	7	78	10	53
St Thomas	5	83	8	61
St John	4	87	7	68
St Lucy	4	91	8	76
St Peter	4	95	7	83
St Joseph	3	98	5	88
St Andrew	2	100	12	100

3 Set up axes of a graph for values 1–100 per cent, with the x-axis showing the cumulative percentage of the population and the y, the cumulative percentage of the area (columns 2 and 4 above).

4 Plot each variable according to its cumulative percentage population (x) and cumulative percentage area (y).

5 Join the points plotted to form a smooth curve. The line $x = y$ which indicates an even distribution between area and population may be indicated.

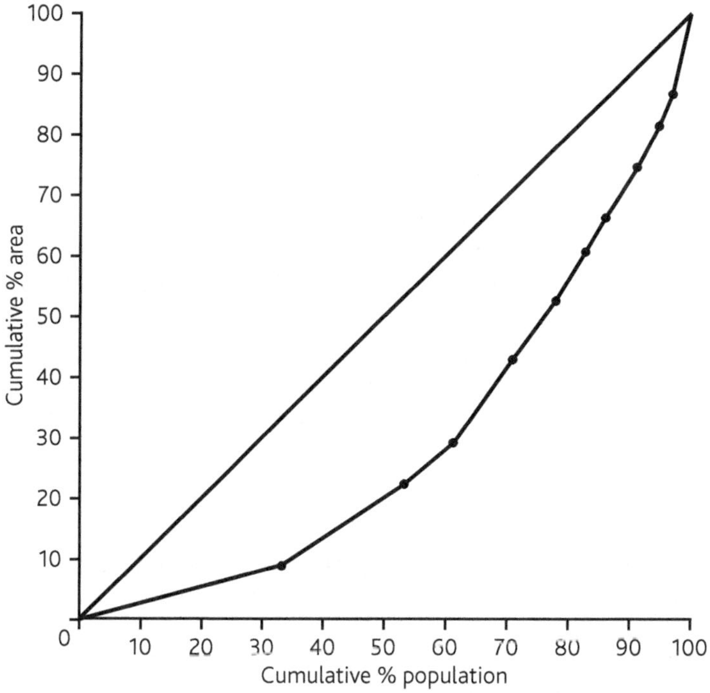

Figure 1.3 *Lorenz curve of population distribution of Barbados (2000)*

You may wish to practise constructing a Lorenz curve by attempting the following activity.

1 Collect the data for population and area for the 10 largest (by population) Caribbean territories.
2 Use the data to draw a Lorenz curve.
3 Describe the population distribution illustrated by the curve.

Advantages of the Lorenz curve for displaying population inequality

- It provides a visual representation of inequality of population distribution.
- It is sufficiently simple that it can be compared across countries and be easily interpreted.

The Lorenz satisfies at least three important principles:

- Scale independence: it does not consider the size of the country, the way it is measured, or whether it is a dense or sparely settled.
- Population independence: it does not matter how large the population of the country is.

Disadvantages of the Lorenz curve

- The Lorenz curve may not give an accurate picture of actual inequalities if the resources of the area are unevenly distributed.
- Populations of similar size may have dissimilar shaped curves.
- It can only be analyzed in terms of cumulative percentages.
- The measure will give different results when applied to different levels of data, for example, constituencies instead of parishes.
- As for all statistics, there may be systematic and random errors in the data.

Key Points

- Population is unevenly distributed in an area, region, country and globally.
- Generally, areas where climatic conditions are not extreme and with abundant natural resources tend to attract larger populations.
- A range of factors that are generally termed physical (environmental) and human influence population distribution.
- Dot maps and the Lorenz curve are often used to illustrate population distribution.
- Using dot maps and the Lorenz curve to depict population distribution result in a graphical representation of population over an area. However, these techniques may lead to an over-simplification of reality.

Conclusion

The study of population distribution remains critical to an understanding of the reasons why people locate the way they do. While many distribution patterns may have had their genesis in historical factors, contemporary forces are also playing an important role. Being able to identify areas of largest population numbers can allow policy-makers to plan more effectively and more strategically. In a world of scarce resources, this is essential to ensure that optimal use of the resources is always the over-riding consideration.

End Test

1 Population distribution is the:
   a spread of people
   b settlement patterns
   c persons per area
   d persons along roads.

2 One disadvantage of the dot map for showing population distribution is:
   a distribution cannot be shown
   b the overlapping of dots in dense areas
   c too few dots on entire map area
   d large dots cannot fit the map.

3 The Lorenz curve is useful for investigating population distribution because it shows the relationship between:
   a cumulative population and resources
   b cumulative area and population
   c food and population
   d density and area.

4 Figure 1.4 below shows a Lorenz curve of population distribution:

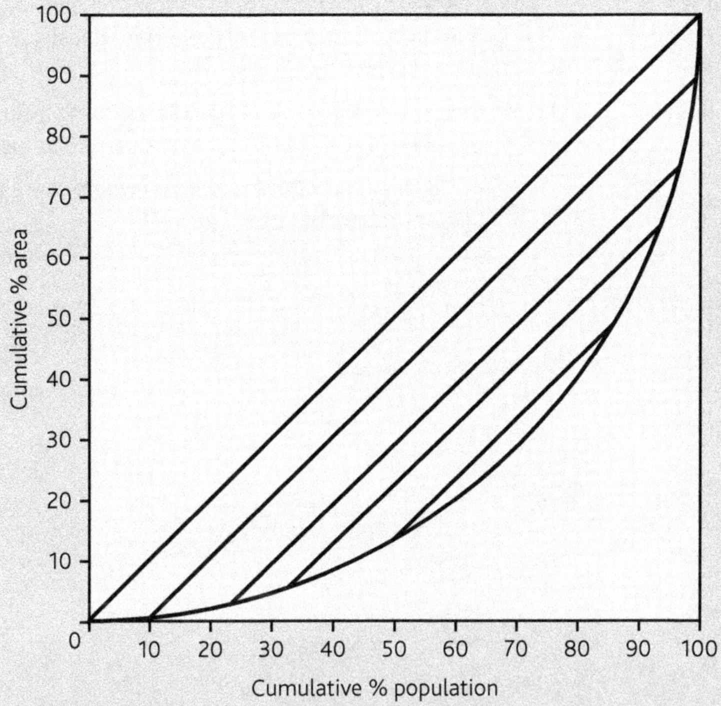

Figure 1.4 *Lorenz curve showing population distribution*

Discuss the information that Figure 1.4 provides on population distribution.

5 Outline TWO human factors that influence variations in population distribution.

6 Discuss TWO advantages and TWO disadvantages of using the dot map to represent population distribution.

End Test Feedback

1 a

2 b

3 b

4 The figure indicates that population is unevenly distributed, with some areas densely populated and others being sparsely populated. For example, 80 per cent of the population occupies approximately 30 per cent of the total land area, while the remaining 70 per cent of land is only occupied by 20 per cent of the population.

5 Economic considerations remain some of the most important factors in determining population distribution. These considerations are often viewed in terms of:

- costs,
- revenue, and/or
- profits/wages.

i For example, transportation costs make it difficult to exploit the resources found in the interior of South America. In general terms, the longer the journey into the interior, the greater the transportation costs and the less profitable economic operations in these parts are likely to be. Population distribution, therefore, in the interior of South America remains low.

ii But the perception of good, rewarding jobs are often a lure for persons to move to a particular area, often in an urban centre. For example, migrants to Mexico City, London, or New York are often under the impression that they will receive better working conditions and financial benefits (salaries, allowances, pension plans) than if they had remained in their current location. The receiving centres of such migrants tend to experience a high concentration of population.

6 Advantages of the dot map:

- It is easy to understand once the concept is understood
- it is effective for showing variations in space of phenomena which exists in relatively large quantities.

Disadvantages of the dot map:

- There is often difficulty in estimating density: human tendency to underestimate density; not used for density
- dots may be interpreted as representing a single instance of the phenomena at a particular location;
- it is difficult to estimate density where there are large numbers of dots in a small area.

2 Population Change – Natural

General objectives

On completion of this module you should:

- understand the factors affecting growth of human populations
- develop an understanding of demographic processes
- develop appropriate skills and techniques in Human Geography.

Specific objectives

You should be able to:

- analyze the components of natural population change
- explain the causes and consequence of natural population change.

Introduction

Most countries undertake a census of their population, generally, every 10 years. Data on the age and sex of population as well as other variables, such as numbers of births, are collected. Data on the number of immigrants and emigrants are also collected. These data provide the necessary information on population change in the country: both by natural increase and net migration (Module 3 Population Change – Migration). On a global scale, however, migration has no effect on population change.

Analysis of natural population growth, natural increase, is important for planning aspects of development of a country. Policies may be developed to address the factors influencing birth rates, death rates and life expectancy.

In some countries population growth outstrips the countries' ability to provide for their needs. These countries try to restrict the birth rates by enforcing anti-natalist policies. Other countries have declining populations and may offer incentives to increase their birth rates by adopting pro-natalist policies.

Content

- Factors influencing birth rate, death rate, natural increase, fertility rate, life expectancy, doubling time (including calculations).
- Demographic Transition Model and its applicability to the experiences of developed and developing countries.
- Population policies – case studies of pro-natalist and anti-natalist policies.

Birth rate

Populations change naturally because of the relationship between numbers of persons being born (the crude birth rate) and numbers dying (the crude death rate).

The crude birth rate refers to the number of live births per 1000 population in a given year. It is crude because the number of births is related to the total population without any consideration of the age and sex composition and a country's birth rate is affected by its age and sex composition. A country with a high proportion of young adults will have a high birth rate. The median age of the population of Mali in West Africa in 2009 was 15.6 and the percentage of the population in the 0 to 14 age group was 42.9. The corresponding figures for Finland were 42 and 17.9. Mali has a birth rate of 49.2 per 1000 and Finland, 10.4. Moreover, given the size of the young population, population growth in Mali may continue to be high. Birth rates of 30 and above are considered to be high. Birth rates of less than 18 per 1000 are considered low. In 2009, Barbados had a birth rate of 12.6, Trinidad and Tobago, 13.2 and Jamaica 20.4 per 1000. Countries with a low proportion of women also have low birth rates.

The birth rate is also influenced by culture and social customs such as religion, the age of marriage, the status of women and by the population policies adopted by countries. In countries where there is opposition to contraception birth rates are generally high. For example, in some countries where a large proportion of the population comprises Roman Catholics or Muslims, birth rates are high. This is a generalization that does not always hold true and there are notable exceptions because of the many factors that affect a country's birth rate. Over 90 per cent of the population of Italy are Roman Catholics but Italy has one of the lowest birth rates (8.2 per 1000) in Europe.

The level of education in the country and especially the level among women affects the birth rate. Where women have access to educational opportunities and are involved in what the International Labour Organization (ILO) describes as 'decent' work, they limit their family size. The decisions of such women who work full time during the child bearing years, are influenced by government policies such as funded health and child care benefits. For example, Scandinavian countries have more generous benefits for families with children than Italy and the birth rate in Sweden is 50 per cent higher than in Italy.

Fertility rate

The birth rate is crude because its denominator contains males as well as females who are not in the child bearing age group. Because of differences in age and sex composition across the world, the birth rate is not useful for the purpose of comparison. Far more useful is the fertility rate which measures reproduction among the child bearing group. The total fertility rate is the average number of children that a woman would bear if she lived her full reproductive lifetime. In a country with a total fertility rate of two, a woman would have, on average, two children during her lifetime. Table 2.1 shows the total fertility rate for selected countries.

Table 2.1 *Total fertility rate – (2012 est.)*

Country	Fertility rate
Italy	1.4
United Kingdom	1.91
USA	2.06
Barbados	1.68
Trinidad and Tobago	1.72
Jamaica	2.21
Mali	6.35

Source: CIA World Factbook

It may seem that if a woman has two children during her lifetime she would be able to replace herself and her husband. However, allowances must be made for those women who do not have children and for infants and other deaths in the population. The replacement fertility rate is considered to be between 2.1 and 2.3 or even higher, depending on the mortality rate.

Death rate

The death or mortality rate is also expressed as deaths per 1000 persons of a population in a year. This is called a 'crude' death rate. Generally, mortality rates are higher in less developed than more developed countries. Table 2.2 shows the death rate for selected countries. From the table it is clear that the countries in the Caribbean have a lower death rate than those in more economically developed countries such as Italy and the United States of America.

Table 2.2 *Death rate per 1000 persons (2012 est.)*

Country	Death rate
Mali	13.94
Italy	9.93
USA	8.39
Barbados	8.39
Trinidad and Tobago	8.35
Guyana	7.18
Jamaica	6.59

Source: CIA World Factbook

However, these rates are not meaningful for comparative purposes because the age structure of these countries differs. Countries with a high proportion of young adults will have a lower death rate than countries with an ageing population. The proportion of the population aged 65 years and over in Italy (20 per cent) is twice that of Barbados. To overcome this problem, the death rate is calculated for different age groups. This is the age specific death rate. The infant mortality rate – deaths in the first year of life – is important because improvements in health and health services have had the most significant impact on these ages. The infant mortality rate in Mali is 102.5 per 1000 live births, the rate in Italy is 5.5.

Death rates are not only dependent on the age structure. In many developing countries death rates have declined because of modern medicine and improvements in sanitation. There have also been improvements in social conditions – better diets, better housing. As a result, life expectancy at birth or the number of years an individual is expected to live or their lifespan is increasing. However, in countries affected by the HIV/AIDS epidemic, life expectancy is falling. Two thirds of all people infected with HIV/AIDS live in sub-Saharan Africa and the virus has affected health, education, livelihoods and the workplace. Life expectancy at birth in Botswana fell from 65 years in 1990–1995 to 40 in 2000–2005. The epidemic is also spreading rapidly in Asia.

Natural increase

The natural increase is the difference between birth rates and death rates. Increases or decreases in the population resulting from migration are not included. For example, the birth rate in Jamaica in 2009 was 20.4 per 1000 and the death rate, 6.4. Its rate of natural increase was 14 per 1000. This rate is usually expressed as a percentage, that is, 1.4 per cent annually. The natural increase in Germany is –0.2 per cent and without immigration its population would be shrinking. When a country's birth plus immigration are equal to deaths plus emigration, that country is experiencing zero population growth. Germany is experiencing zero population growth. The population of Japan is declining.

It is useful to know how quickly a population is growing. One method is to calculate it's doubling time. This is the number of years it takes a population to double itself if the present growth is maintained. This can be roughly estimated by dividing 70, the natural logarithm of 2, by the percentage growth rate:

$$t = 70/k \text{ (growth rate as a percentage)}$$

Jamaica with a growth rate of 1.4 per cent will have a doubling time of 70/1.4 = 50 years.

Anti natalist policies

A belief that a rapid population growth would retard development has caused several countries to implement policies to restrict growth. China's polices have been the most successful. In 1965, the population was 540 million and the birth rate 37 per 1,000. The one child policy was introduced in 1979 with exemptions for ethnic minorities. Late marriages were encouraged, free contraceptives made available and the programme was backed up by incentives as well as penalties.

Incentives included:

- cash awards
- preferential access to housing, education and health services.

Penalties included:

- fines
- job loss
- sterilization of husband or wife.

In 2009, the birth rate in China was 14 per 1,000. It has been estimated that the policy has averted about 400 million births. But the policy has created several problems:

- The '4-2-1' problem which summarizes the strain on one child to provide for four grand parents and two parents.

- Gender imbalance. Sons are preferred and couples employ illegal means to ensure that their one child is a son. Female foetuses are sometimes aborted or girls abandoned. There are now 120 males for every 100 females. It is estimated that there are between 40 to 60 million 'missing' females.

Pro natalist policies

There was a decline in fertility in France in the 1960s as in many European countries. France faced a labour shortage and a demand for women to work in the services. Policies were implemented to boost the birth rate. France's fertility rate (1.9) is now the second highest in Europe. Incentives include:

- Community funded day care facilities and nursery schools.
- Family allowances to families with at least two children.
- Allowances for children with handicaps: in lone parent families.
- Allowances to cover costs associated with the new school year.
- Child allowances for children less than three years old.
- 16 weeks maternity leave, with 26 weeks for the third child.
- Paternity leave of two weeks.

Table 2.3 *Infant mortality and life expectancy for selected countries (2012 est.)*

Country	Infant Mortality (per 1,000)	Life Expectancy (per 1,000)	Country	Infant Mortality (per 1,000)	Life Expectancy (per 1,000)
Angola	83.5	54.6	Japan	2.2	83.9
Australia	4.6	81.9	Mozambique	76.9	52.0
Bangladesh	49.0	70.1	Nigeria	74.5	52.0
France	3.4	81.5	United Kingdom	4.56	80.2
India	46.1	67.1	United States of America	6.0	78.5

Source: CIA World Factbook

Activity 2.1

Investigate the concept of 'decent work'. Why would women who have access to 'decent work' limit their family size?

Use the data above to answer the following questions:

1 Describe the main patterns of
  a infant mortality rates and
  b life expectancy figures.

2 Outline three reasons for the patterns you have identified in (1a).

Feedback

1 a Mortality rates are declining in developed countries but infant mortality rates remain high. High rates are recorded in some countries in Africa and Asia; while the lowest rates are in Japan, Europe and North America.

  b The main pattern of life expectancy is very similar: developing countries have low life expectancies, for example, Angola; while people in the developed countries, such as Japan, can expect to live into their 80s.

2 Three reasons which account for the patterns of infant mortality in 1. a) above are nutrition, medical and disease conditions in those countries. In the developing countries, multiple births and poor diet may result in low birth weight and sickly babies who die before they are one year old. While in the developed world fewer babies are born to very healthy parents, so babies survive past year one. Medical conditions for pregnant women and babies in the developing countries are often very inadequate. Babies may be born in very unsanitary conditions with little medical assistance available. In the developed countries, pre and ante natal

medical supervision is the norm so the development of the foetus and baby are closely monitored. Young babies are very susceptible to disease. In developing countries gastroenteritis and other water borne diseases can kill babies who are already malnourished. In developed countries, many of these early childhood diseases have been eradicated.

The Demographic Transition Model

Models are generalizations that present an idealized picture of a phenomenon. The Demographic Transition Model (DTM) was developed to demonstrate the transition from high birth and death rates to low birth and death rates with economic development. The model was based on experiences of the developed countries during industrialization and applied to these countries as a group. The original model proposed four stages in the transition. These stages were:

- Stage 1 – High fluctuating (pre modern)
- Stage 2 – Early expanding (urbanizing, industrializing)
- Stage 3 – Late expanding (mature industrial)
- Stage 4 – Low fluctuating (post industrial)

A fifth stage is sometimes added to take account of developments in countries where fertility is at replacement level (Figure 2.1).

The applicability of DTM to population change in developed and developing countries

The DTM is very easily applied to developed countries. Generally, they experienced the changes in birth and death rates proposed by the model as they industrialized their economies. It therefore, shows its best fit with industrialized nations of Europe and North America. Similar cultural norms and improved status of women in the 20th century in these areas contributed to the applicability of the model. However, a fifth stage is proposed to take into account declining populations not perceived by the model.

In the developing world the model does not accurately map the stages of population change. A significant difference is the economic conditions which see many countries struggling to industrialize in the late 20th and 21st centuries. Many developing countries experience more rapid population change with death rates falling sharply. They have also experienced higher maximum rates of growth. There has been a longer lag in developing countries between falling death and falling birth rates. Fertility changes require changes of behaviour that sometimes come into conflict with cultural values.

However, in both regions, infant mortality was high in the period when total fertility was high and fell with declining fertility.

Key Points

- Populations are very dynamic in time and space.
- Populations grow or decline by natural increase and net migration.
- The rate of natural increase is influenced by population structure as well as social and economic factors.
- The Demographic Transition Model can be applied to analyze population change with economic development.

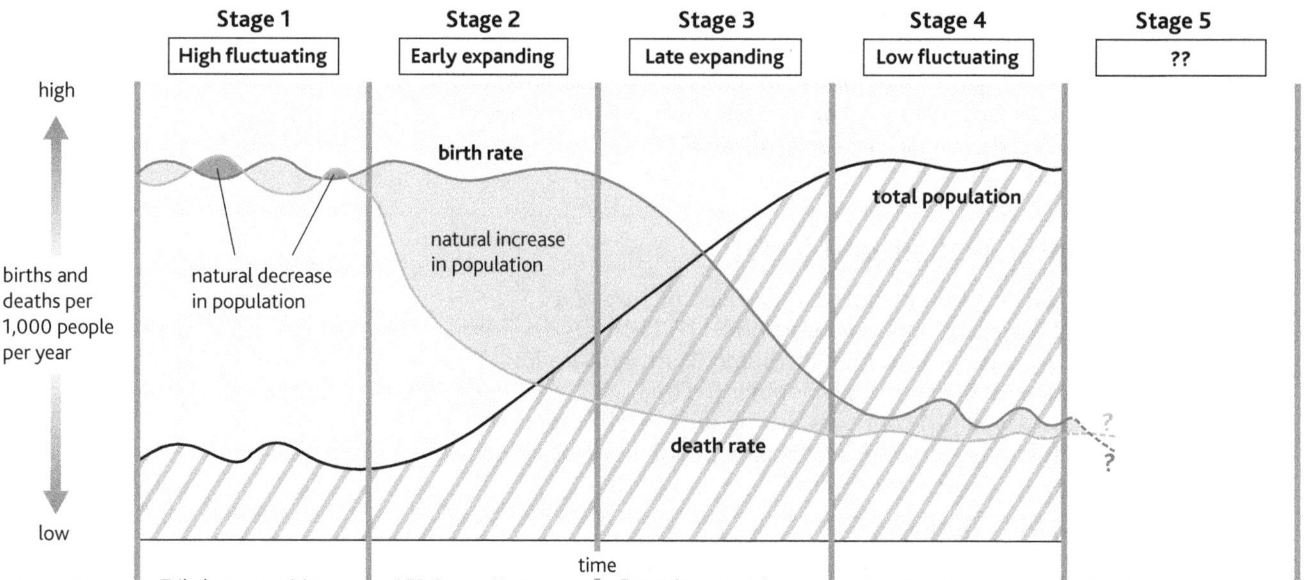

	Stage 1	Stage 2	Stage 3	Stage 4	Stage 5
present world examples	Tribal communities (UK pre-1760)	LEDCs, e.g. Kenya, Ethiopia, Bangladesh (UK 1760–1880)	Emerging countries e.g. China, Brazil, India (UK 1880–1940)	MEDCs e.g. Japan, USA, Argentina (UK post 1940)	Zero growth, e.g. Sweden, Italy
	Stage 1: Here both birth rates and death rates fluctuate at a high level (about 35 per 1,000) giving a small population growth.	**Stage 2:** Birth rates remain high, but death rates fall rapidly to about 20 per 1,000 people giving a rapid population growth.	**Stage 3:** Birth rates now fall rapidly, to perhaps 20 per 1,000 people, while death rates continue to fall slightly (15 per 1,000 people) to give a slowly increasing population.	**Stage 4:** Both birth rates (16 per 1,000) and death rates (12 per 1,000) remain low, fluctuating slightly to give a steady population.	(Will there be a **Stage 5** where birth rates fall below death rates to give a declining population? Evidence suggests that this is occurring in several Western European countries although growth rates here are augmented by immigration.)

Figure 2.1 *Model of population change*

Activity 2.2

1 Describe the main characteristics of Stage 4 of the model.

2 At what stage of the model would most countries of the Caribbean be placed?

3 Define the term 'natural increase'.

Feedback

1 The main characteristics of Stage 4 are low and fluctuating birth and death rates and small population growth.

2 Most countries of the Caribbean are at Stage 3.

3 Natural increase refers to the difference between the crude birth and death rates.

Conclusion

Population change is significant over time and space. Demography investigates population growth and decline in different parts of the world. Calculations of population growth help to express these changes. Some countries strive to reduce their populations while others seek to increase them.

End Test

1 Explain the term 'fertility rate'.

2 In some developing countries, fertility rates remain high. With reference to conditions in such countries, outline factors which help to explain these high rates.

End Test Feedback

1 Fertility rate is the number of live births per 1,000 from women of child bearing age 15–49 in a given year. Generally, the fertility rate in less developed countries (LDCs) is higher than in more developed countries (MDCs).

2 There are a number of factors which can help to explain high fertility rates in some countries. One of the most significant factors is the status of women. In some parts of Africa tradition demands high rates of reproduction. There is an expectation that a woman should bear as many children as possible. In addition, lack of education and knowledge of contraception can result in continuing high fertility rates. Some men may not allow their wives to use any contraception in the belief that her fertility enhances his maleness.

In other countries especially in the Middle East and Asia, religious restrictions create high fertility rates. Roman Catholic and Muslim religions are opposed to artificial birth control and this may result in large families.

Countries with high infant mortality rates also have high fertility rates as a woman must have many children to ensure the survival of a few. This is important in countries where children are seen as economic assets. They are seen more as potential workers than more mouths to feed.

3 Population Change – Migration

General objectives

On completion of this module you should:

- understand the factors affecting migration of human populations
- develop an understanding of demographic processes
- develop appropriate skills and techniques in Human Geography.

Specific objectives

You should be able to:

- analyze the components of migration population change
- explain the causes and consequence of migration population change.

Introduction

Migration is movement from one place to another usually across a political boundary. It may be permanent or semi permanent and such movements have been taking place throughout human history. These movements can have significant consequences for population change of towns, cities and nations. Large migrant streams can alter population composition, as in the case of past African migrations to the New World, as well as structures.

People migrate for a variety of reasons. They may leave countries which do not offer employment opportunities; in which the quality of life is low; because they fear for the security of themselves or their families; to escape the ravages of war and intolerance. Some are also forcibly moved from their homes.

Content

- Types of migration (including construction of flow lines).
- Causes and consequences of migration.

Types of Migration

Migrations can be voluntary or forced. Voluntary and forced migration can be subdivided into internal, external (international), temporary and permanent.

Forced migration

Forced migration is unintended movement resulting from conflict, development policies, projects and disasters. People may be forcibly removed from their homes and relocated to make room for economic developments. Millions of people today are forced to leave their homes but still reside within the borders of their own country. A good example of such forced internal movement was that associated with the building of the Three Gorges Dam in China. The dam was built across the Yangtze River to control flooding and provide hydroelectric power to keep pace with China's economic development. Hundreds of villages and towns were inundated and over 1 million people forced to leave the Chongqing Municipality for infertile mountainous land along the river. Worldwide, internally displaced persons (IDP) were thought to number about 25 million in 2003. More than a half of them are in Africa – Sudan, The Democratic Republic of Congo (DRC), Uganda and Angola. There were about 2.8 million IDPs in Iraq in 2008, many of whom had fled from Bagdad to safer zones in the far north, south and west of the country. Large numbers are also found in Sri Lanka and Colombia.

The slave trade between Africa and the New World was an example of forced external migration. The partition of India and Pakistan forced large scale movement across the new international borders. Today, there are large numbers of forced migrants (refugees) who have had to flee war, violence and chaos in their homelands – the mass movement from Rwanda in the 1990s to escape genocide; the migration caused by conflicts in Iraq, Georgia, Armenia and Yugoslavia. Human trafficking is regarded as contemporary slavery and forced labour involving domestic servitude, child soldiers, agriculture and the commercial sex trade. The volcanic eruptions in Montserrat forced many to leave their homes for foreign countries. Climate change is expected to produce new waves of forced migration and it is estimated that by the year 2050, about 200 million persons may be displaced by climate change.

Some of the forced migrants settle permanently in their new homes. Others return home when the threats have passed.

Voluntary migration

This is the free movement of individuals or groups from their homes. Voluntary migration may be internal and such movements may be temporary (daily commuters) or permanent relocation. Commuters between home and work are temporary internal migrants (a movement sometimes referred to as 'circulation'). Millions of workers in China leave their homes in rural areas to work in the cities. Central Business Districts of cities that teem with pedestrians during the day are often deserted by night; while suburbs are empty by day but filled by night.

Internal migration in Barbados

Table 3.1 *Barbados population by Parish 1960–2000*

Year	1960	1970	1980	1990	2000
Christ Church	33,425	35,987	40,206	46,943	49,497
Saint Andrew	7,813	7,163	6,720	5,624	5,254
Saint George	17,255	16,841	17,190	16,718	17,868
Saint James	13,611	13,379	16,941	20,771	22,741
Saint John	10,967	10,524	10,256	9,640	8,873
Saint Joseph	8,582	7,849	7,213	7,204	6,805
Saint Lucy	8,997	8,746	9,239	9,277	9,328
Saint Michael	93,716	97,567	96,916	89,840	83,684
Saint Peter	10,860	10,622	10,623	10,055	10,699
Saint Philip	17,075	16,810	18,337	20,540	22,864
Saint Thomas	10,026	10,539	10,587	10,676	12,397
11 parishes	**232,327**	**236,029**	**244,228**	**247,288**	**250,010**

Source: National Census Report 2000, Barbados

Over the 40 year period, parishes of St James and Christ Church have gained most people (Table 3.1) and grew by 67 per cent and 48 per cent, respectively. The rural parishes of St Andrew, St Joseph and St John had the biggest losses –33 per cent, –21 per cent and –19 per cent, respectively, in the period. The parish of St Michael with the capital Bridgetown lost 11 per cent of its population. These movements demonstrate the pull of economic activity, of tourism on the south and west coasts but also the attraction of the newer business or commercial centre at Warrens on the boundary between St James and St Michael. The decline in St Michael demonstrates suburbanization, with people moving out of the crowded urban area to the suburbs. The decline of sugar cane cultivation and demand for housing accounts for the significant growth in St Thomas (24 per cent) and St Philip (34 per cent). St George and St Lucy with the smallest growth of 4 per cent in the remaining agricultural areas. These are permanent movements.

Internal movements may be rural to urban and this is more common in less developed countries. The pattern in more developed countries is typically urban to rural associated with suburbanization and counter urbanization (see Module 6). Some also move within rural and within urban areas. Internal movements usually occur over comparatively short distances.

External/International. The large scale movement from the Caribbean to Britain in the 1960s; to USA and Canada are examples of external or international migration. Many settled permanently in these countries and they and their descendants form large communities in many urban areas in developed countries.

There are also temporary international migrants. Labour migrants such as Caribbean farm workers who go to Canada and the United States of America to harvest crops, return at the end of their contracts. They are temporary international migrants. The global recession in 2009 forced many migrant labourers to return home. For example, 72, 000 Bangladesh migrant labourers returned home from Saudi Arabia and Dubai. Tourists are also temporary migrants leaving their homes to seek leisure and relaxation. But they are fully intent on returning to their place of residence. Students who pursue their studies in other countries and return after the course of study are also temporary migrants.

Migration patterns

Spatially, there are several migration patterns:

Step migration in which the migrants change location in a series of small steps, each taking them closer to their intended destination. They may move from a small farm to a nearby town, then to a city and finally to a foreign country.

Chain migration is a series which may begin with one family member who sends money to bring family members to the new location. This process leads to the clustering of people from specific locations in neighbourhoods of the towns in which they settle. These clusters are referred to as migration fields. Eventually migration may be channelized as people move between areas that are connected by past migration.

Activity 3.1

1 Indicate the nature of the following migrations by ticking the appropriate columns.

Migration	International	Internal	Voluntary	Forced
Rwandans to the DRC				
Chinese from Chongqing to Yangtze hills				
Caribbean farm workers to Canada				
Iraqis in 2004 to southern provinces				
Barbadians from the parish of St Andrew to Christ Church				

2 Identify three different types of internal migration.

Feedback

1

Migration	International	Internal	Voluntary	Forced
Rwandans to the DRC	✓			✓
Chinese from Chongqing to Yangtze hills		✓		✓
Caribbean farm workers to Canada	✓		✓	
Iraqis in 2004 to southern provinces		✓		✓
Barbadians from the parish of St Andrew to Christ Church		✓	✓	

2 urban-rural, rural-urban, rural-rural.

Causes and Consequences to Source and Host Countries

The causes of migrations are varied but may reflect a combination of forces in the home and in the destination countries. They are usually classified as push and pull factors. People leave their homes because of a lack of economic opportunity or security and are drawn towards other countries by the perception of opportunity.

The consequences of migration can be both positive and negative. For the sending country, loss of population can have both negative (loss of skilled labour) and positive (relieving population pressure) effects. Similarly the receiving country can be enriched by the culture of migrant labourers but conflicts may arise between the host and migrant society.

Causes and consequences of migration – a sending country, for example, Jamaica.

Causes

Economic

- High unemployment as a result of decline in bauxite and agriculture;
- Low salaries for skilled workers, for example, nurses and teachers. Teachers and nurses recruited for work in developed countries.

Social

- Poor housing/sanitary conditions;
- Inadequate access to health and higher education facilities;
- Crime affecting the quality of life especially in urban areas.

Political

- Discrimination – party political;
- Civil unrest.

Physical

- Natural disasters.

Consequences

- Remittances – 17 per cent of GDP in 2008;
- Brain drain – loss of skilled and professional workers;
- 'Barrel children' – family livelihood depends on receipt of goods shipped from overseas;
- Criminals 'hardened' in life of crime abroad reintroduced to the society;
- Retired returning national add skills and finances.

Causes and consequences of in-migration, that is, to a receiving country, for example, Canada.

Causes

Economic

- Demand for skilled workers throughout country.

Political

- Liberal immigration policy – net migration gain 5.6 persons per 1000.

Cultural

- Tolerance of cultural differences.

Consequences

- Economic prosperity due to increased use of resources.
- Cultural diversity. Culture enriched by presence of immigrants from several countries.
- Conflict – cultural tensions, for example, West Coast between Chinese immigrants and local population. Immigrant customs rejected by host country, for example, forced marriages in Asian families, discriminatory practices against women. Young immigrant males targeted by gangs.
- Concentration of population in urban areas.

Construction of Flow Lines

Flow lines are lines of varying thickness with arrowheads at one end. They are useful for showing volume and direction of movements. When appropriately placed on maps, flow lines give a good visual representation of migrations allowing volume as well as the sending and receiving countries to be identified.

Steps in the construction of flow lines

1. Collect relevant data on migration in country or region.
2. Choose a base map of country or countries on a scale large enough to draw flow lines.
3. Examine the data and determine smallest and largest volume and choose a scale for thickness of line, for example, 1mm represents 1000 persons.
4. Pencil in the direction of the arrows to ensure that they do not overlap or cross each other.
5. Determine the thickness of each line to show the volume of the movement. [It is easier to maintain the desired thickness of a straight line than a curved one.]

6 Place the scale on the map.

7 Draw in the lines of the desired thickness, [these may be shaded].

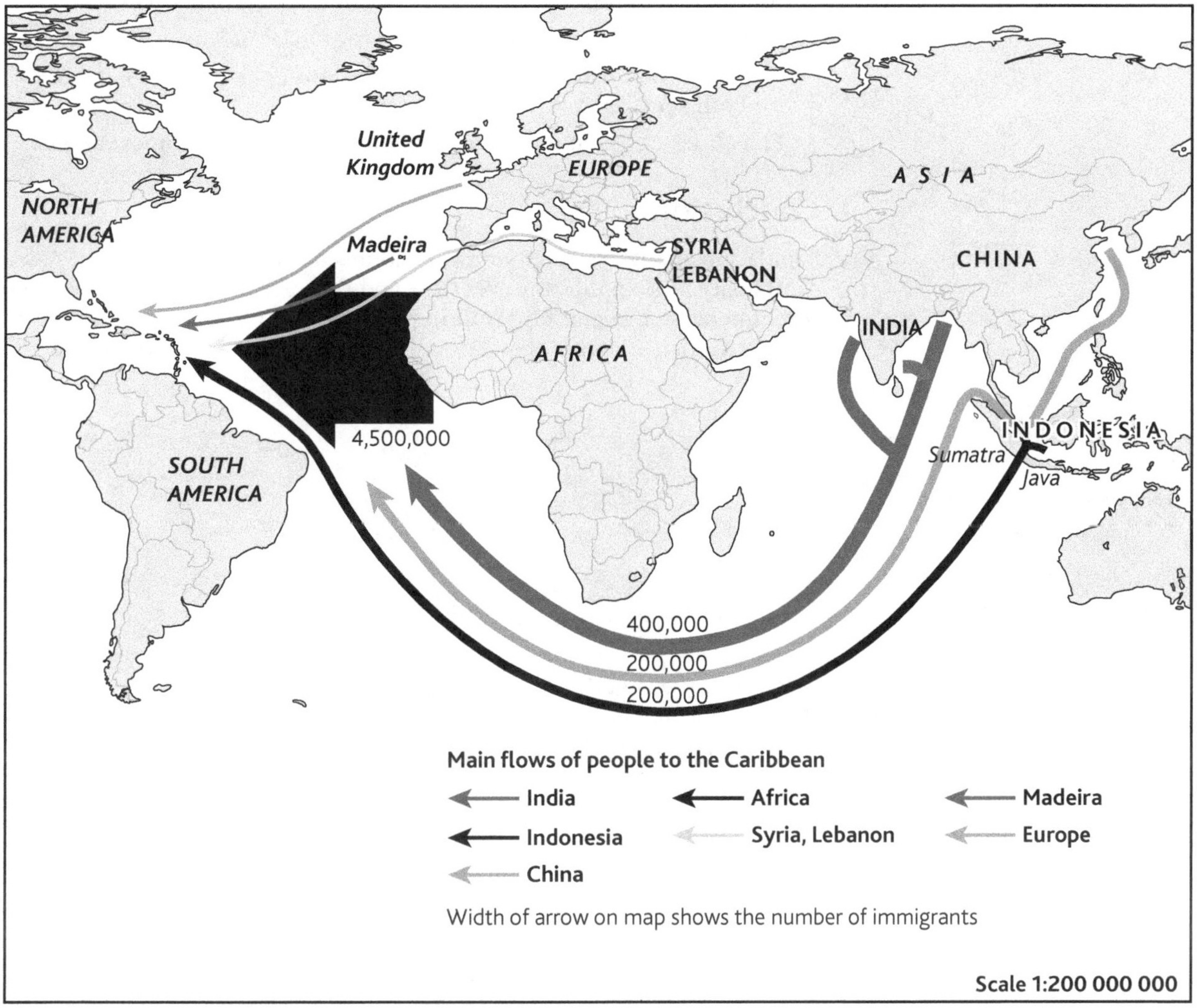

Source: Atlas for Caribbean Examinations. 3rd ed. Pearson Education Ltd. 2007

Figure 3.1 *Migration to the Caribbean 1500s to 1920s*

Activity 3.2

1 Measure the width of the flow line representing 200,000 persons and, therefore, determine the scale of the flow lines.

2 Which continent sent the smallest number of people to the Caribbean?

3 How many migrants came from China?

Feedback

1 The line is 1mm thick and represents 200,000 persons. Therefore, the scale of the flow lines is 1 to 200,000.

2 Europe.

3 200,000 persons.

Key Points

- Populations move around within (internal migration) and over (international migration) national boundaries for many reasons.
- Large numbers of people today are being forced to leave their homes.
- Migration has significant consequences, both positive and negative, for both sending and receiving countries.
- Flow lines can help to identify and quantify population movements.

Conclusion

Population change by migration contributes to overall demographic complexity. It is significant over time and space. The causes and consequences of migration result in population variation over the earth's surface. Construction of flow lines of migration help to identify these changes.

End Test

1 Define each of the following terms:
   i involuntary migration
   ii circulation
   iii step migration
   iv international migration.

2 With reference to specific examples, state two social problems in the destination country that are associated with international migration.

3 State two benefits which countries of origin obtain from international migration.

End Test Feedback

1 i Involuntary migration is forced movement from one area to another, for example, slave trade.
   ii Circulation is the daily movement from home to work, for example, commuting.
   iii Step migration is serial movement from settlement to settlement until ultimate destination is reached, for example, Jamaica – Mandeville-Spanish Town-Kingston.
   iv International migration is movement across country borders. Most countries try to control the numbers of persons moving across their boundaries both immigrants and emigrants.

2 Two social problems associated with international migration in receiving countries are conflict over status of women and ethnic violence. In many host countries, women are permitted and encouraged to participate fully in social life, however many Asian migrant women are not permitted to associate socially with non-family males. This has lead to murder and restrictions on migrant females which are illegal in the host country. For example, there are runaway Muslim girls in the United States of America. Many Asian migrants who are ethnically different from the host population perceive themselves as being unjustly treated. They may target and be targeted for inter gang violence. For example, in the United Kingdom the white 'skinhead' neo-Nazis and the black inner city gangs are often victims of assault.

3 Remittance flows which formed 17 per cent of the GDP of Jamaica in 2008.

   Returning migrants bring the skills and sometimes the capital which they acquired in the developed country.

4 Population Structure

General objective

On completion of this module you should develop an understanding of demographic processes.

Specific objective

You should be able to analyze the components of population structure.

Introduction

Structure refers to the age and sex balance in a population. It can be displayed in a population or age-sex pyramid. This is a combination bar graph that shows the number or the percentage of the population by age and sex. The shape of the pyramid reflects the long and short term trend in population growth both by natural increase and net migration.

Population pyramids can be used to determine which stage a country is on the DTM which was examined in Module 2. The pyramids vary from the short broad-based shapes of youthful populations in LDCs to the long, narrow, top-heavy shapes of aging populations in MDCs. The relationship between the sizes of different age groups in a population can indicate the dependency ratio which has an impact on the economic development of a country.

Content

- Population structure in LDCs and MDCs.
- Construction and interpretation of population pyramids.
- Interpretation of population structure in LDCs and MDCs.
- Dependency ratios – calculation and implications.
- Case studies of youthful and aging populations.

Population Structure in LDCs and MDCs

Population structure in LDCs

Generally, the population structure of LDCs is marked by a higher proportion of young persons, with a sharp decline in numbers at increasing ages. These pyramids have wide bases because of high birth rates and narrow sharply because of high death rates. Life expectancy may be as low as 50 years in some LDCs. There may be sex inequalities in the population structure depending on migration patterns. Some may be 'sending' countries, for example, Guyana and Jamaica. Therefore they may have 'missing' males or females of working age. Wars and high maternal mortality also result in sex inequalities. This structure is sometimes referred to as a 'youthful' one.

There is a great variety of population structures within the LDCs. At one extreme, poorer countries such as Papua New Guinea have very young populations while more developed LDCs, like Barbados have aging populations. Differences in culture, religion and status of women, along with levels of poverty, account for these differences in population structure. HIV/AIDS in Sub-Saharan Africa has reduced some life expectancy in some countries to 40–50 years.

Most Caribbean countries would be considered as having 'youthful' populations, although there is some variation among countries.

Population structure in MDCs

The population structure in more developed countries tends to show a narrow base reflecting lower birth rates and a relatively small proportion of its population under age 15. The birth rate is very low, sometimes below replacement level. The apex of the pyramids tends to be broader than those of less developed countries because life expectancy is high. Medical care is available from cradle to grave and ensures low infant mortality rates and long lives. The crude death rate may be high as a result of an ageing population.

There are variations among MDCs. Countries, such as Ireland have higher BR than Sweden with its declining population. MDCs, such as the USA and Germany with high immigrant populations and have more complex population structures. For example, males may outnumber females in the working age group as this is typically the migrant group. Migrants may have higher birth rates than the resident population as a result of cultural differences.

In both LDCs and MDCs, women outlive men, so the population structure after age 60 is almost always imbalanced in favour of females.

Construction and Interpretation of Population Pyramids

Population pyramids are the main tools for representing population structures. The pyramid consists of two sets of adjacent horizontal bars representing the male and female population by age group (the data may be numerical or percentages). The declining length of bars with increasing age gives the typical pyramidal shape – wide base to pointed apex. Although so named, the actual shapes evidenced by the population structure of different countries may even be an inverted 'pyramid' showing a narrow base and wider apex. The shape of the population pyramid may be used to place the country in one of the stages of DTM (Module 2 – Population Change – Natural), although it should be noted that unlike the DTM, population pyramids include the net migrant population well as natural increase.

Construction of population pyramids

Collect population data by age and sex as given below.

Table 4.1 *Population data for country P*

Age Group	Less than 5	5–9	10–14	15–19	20–24	25–29	30–34	35–39	40–44	45–49	50–54	55–59	60–64	65–69	Greater than 69	Total
Male (thousands)	1,845	1,977	1,980	1,900	1,804	2,118	2,374	2,350	2,030	1,883	1,993	1,552	1,389	1,218	2,603	29,016
Female (thousands)	1,753	1,897	1,872	1,797	1,716	2,045	2,318	2,292	2,006	1,884	2,005	1,589	1,452	1,349	4,101	30,076

Choose suitable scales: horizontally (A) and vertically (B). (The letters refer to the completed pyramid Figure 4.1.)

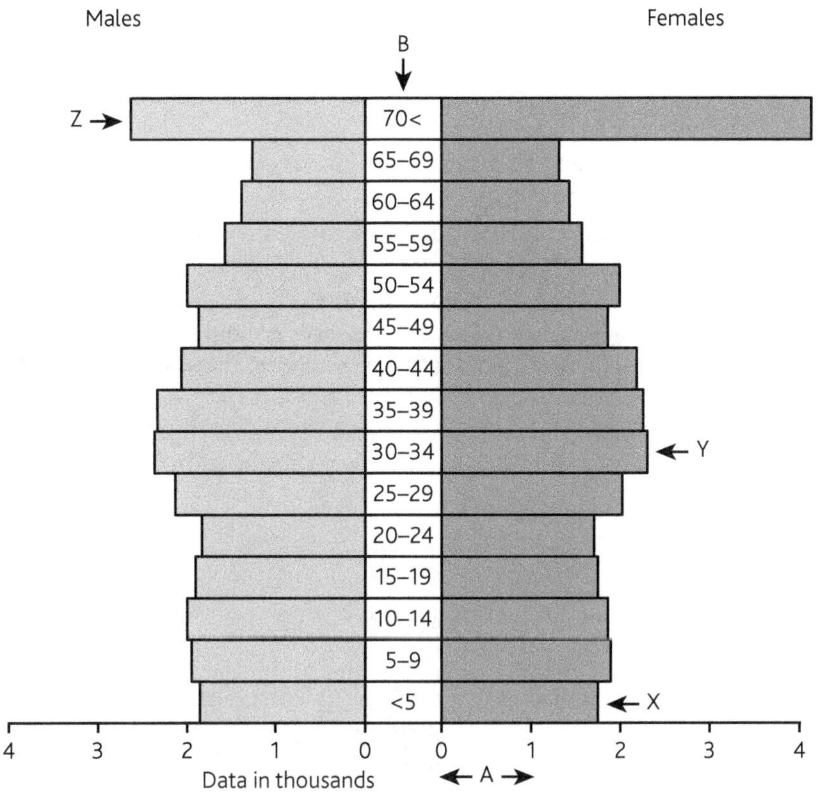

Figure 4.1 *Population structure for country P*

- Horizontally the largest total to be represented is females 4101, and males 2603. The total is 7000. The width of paper available is 8.5cm therefore let 2cm represent 1 million (thousands of thousands). Draw and label the scale leaving a space in the middle for age groups as A on Figure 4.1. Note: The scale must be labeled in thousands ('000s) as the data are given in thousands. (Note: Data could be given as percentage of population.)
- Vertically, there are 15 age groups and the paper is >20cm long so a scale of 1cm represents each age group would fit and result in large, easily readable bars. This is B on Figure 4.1. (Age groups could be in 5 year intervals.)

Carefully plot the data by drawing bars of the appropriate length as indicated by the scale. Conventionally the data for males are drawn to the left and females to the right.

Interpretation of population pyramids

There are four critical aspects of the pyramid:

1 *Shape.* This refers to the overall slope created by the shortening bars. It may be:
   - concave that is, declining rapidly with age;
   - even steep slope – true pyramid-shape;
   - straight especially at base – shortening slowly;
   - convex – bulging in the middle and to top;
   - inverted – wider towards the top.

2 *Base to height ratio*. Generally the base reflects the birth rate and the height the life expectancy.
- Wide base, short height indicate high birth rate and low life expectancy;
- narrowing base and lengthening height reveal falling birth rate and longer life expectancy;
- narrow base and tall height may indicate a very low birth rate and very long life expectancy.

3 *Relative size of 15–65 age groups*. This is the economically active part of the population. Its size relative to the over 65 and under 15 would indicate the size of the dependency ratio, that is, those who have to be supported. It can also suggest the future BR.

4 *Differences in male and female population by age group*. Males may be missing (shorter bars) as a result of war or migration. Females may die in child birth.

Activity 4.1

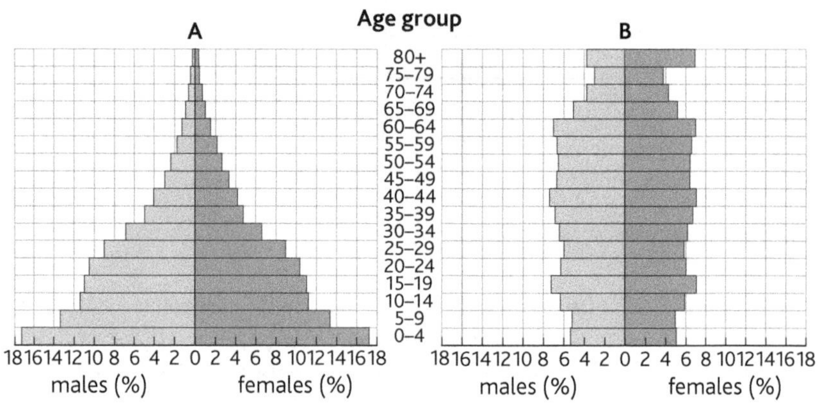

Figure 4.2 *Population pyramids A and B of selected countries (in millions)*

1 Describe the population structure shown by each of the pyramids in Figure 4.2.

2 State whether each is a pyramid of a MDC or LDC and name an example of a country which could have the shape described in 1.

3 Name the stage of the DTM to which each country could belong.

Feedback

1 Pyramid A shows a short concave shape. It has a large number of persons below age 15 and very few over age 65. There is a sharp decline in each age group showing a higher than normal death rate or emigration.

Pyramid B has a much straighter shape; the bars up to age 65 are generally the same length, except those of age 35. The under 15 age group is small indicating a low birth rate. The over 65 group is large with significant numbers over 80 years old reflecting a low DR and long life expectancy.

2 Pyramid A is that of a LDC, a country such as Kenya would have such a population structure.

Pyramid B is that of a MDC, for example United Kingdom.

3 Pyramid A would be in Stage 1 of the DTM. In this stage population growth is 'high fluctuating' showing both high BR and high DR, with low life expectancy.

Pyramid B would be Stage 4 'low fluctuating'. Both BR and DR are low and population growth is low.

Accounting for the characteristics

The population pyramid reflects past conditions. For example in Figure 4.1 country P has a low birth rate hence the narrow base X. But the slightly longer bar Y of the 30–34 age group relates to conditions 30–35 years ago when the country experienced its largest population growth either through immigration or because of a baby boom. At bar Z there is a male deficit in the over 69 age group. This is the age groups born before 1930 that would have been affected by World War I in 1914 and its aftermath. Possibly many men would have died at war and BR would have been low.

The shape of the pyramid reflects not only current conditions in the country but those of the past 85 years.

Interpretation of Population Structure LDCs (Youthful) and MDCs (Aging) Populations

Implications for the future

The population pyramid also allows predictions to be made about the future. The present size of any age group can be used to predict the future size of older groups.

LDCs

In Figure 4.2 A, the broad based pyramid of a LDC such as Kenya, indicates that there is a large pre reproductive group (0–14). The 0–5 age group is larger than any other. One can expect continued growth of the population. This means government would have to plan for reproductive health education and possibly family planning measures to deal with a high fertility rate. There will also be a high demand for jobs. There will be a need for more schools and youth care facilities. Improved medical care may result in lower infant mortality.

MDCs

In Figure 4.2 B, the pre reproductive group is declining and population growth will soon begin to decline. The country experienced its greatest growth 35 to 40 years ago and as this group moves through the population policies will have to be put in place for an aging population. The numbers forming the economically active group have started to shrink. Plans will have to be made for retirement and pensions, care for the elderly and geriatric health facilities. This country would have large numbers of dependant elderly persons especially women, with a decline in the economically active age group. This country may have to import its labour.

Dependency Ratio

The dependency ratio is a measure of the proportion of the population that is dependent either because they are too young or too old to work. It is a hypothetical construct allowing country comparisons and planning implications to be derived. For example terms like 'economically active' are used for working persons, but without reference to actual labour force or employment rates. It is based solely on age characteristics.

The formula for the calculation of the dependency ratio is:

$$\frac{\text{number of persons (under 15 + over 65)}}{\text{number of persons between 16–65}} \times 100$$

For example, data given for country P, dependency ratio =

20,595 [non-economically active or dependent]

divided by

38,497 [economically active] × 100

= 53.5%

A low dependency ratio means that the country will be better able to benefit from its economic production without having to divert resources to support those who are not economically active.

The dependency ratio can be caused by a large youthful population or large elderly population. The ratio can be calculated for each one.

A high dependency ratio will result in a diversion of resources from development projects to support the dependent population.

Activity 4.2

Use the data below to calculate the (i) youth dependency ratio and (ii) dependency ratio.

Table 4.2 *Population in millions*

Age Group	Population (m)
0–4	3.9
5–9	4.1
10–14	3.7
15–19	3.35
20–24	3.65
25–29	3.2
30–34	2.9
35–39	2.8
40–44	2.9
45–49	3.1
50–54	2.9
55–59	2.95
60–64	2.85
65–69	2.5
70–74	1.85
75–79	1.25
80+	1.25

Feedback

i The youth dependency ratio is the total number of persons 14 and under divided by the total number of persons 15–64 expressed as a percentage.

(3.9 + 4.1 + 3.7) m 14 and under

(3.35 + 3.65 + 3.2 + 2.9 + 2.8 + 2.9 + 3.1 + 2.9 + 2.95 + 2.85)m between 15–64

= 11.7/30.6

= 38.24% this is the youth dependency ratio which is moderate.

ii Dependency ratio is the total number of persons 14 and under + total number of persons 65 and over divided by the number of persons 15–65 and expressed as a percentage.

11.7 + (2.5 + 1.85 + 1.25 + 1.25)/ 30.6

= 18.55 / 30.6

= 60.62%

This is a moderate dependency ratio.

Key Points

- Population structure, that is, age and sex characteristics, reflects many of the social and economic conditions in a country.
- The shape of the population pyramid can be used to describe and explain the structure of a country's population.
- The pyramid can be used to predict changes and implications for social and economic planning.
- The dependency ratio is an indication of the demands which may be made on the economic and social resources of the country.

Conclusion

Population structure refers to the age and sex characteristics of a population. It is displayed in population or age-sex pyramids. Population structure reflects the past and has implications for the future.

Tutor Marked Assignment

Figure 4.3 below shows age-sex pyramids, A and B, for two contrasting countries.

a Identify THREE main features of each pyramid.

b Suggest FOUR reasons which may account for the shape of each pyramid.

c Describe TWO social and TWO economic implications for a country similar to the one depicted in pyramid A.

Figure 4.3 *Age-sex pyramids*

Feedback

a Pyramid A

- Wide base and/or more than 25 per cent under 15 years
- High percentage in 20–29 age group
- Small percentage in 65 and over age group
- Excess males in the 20–29 age group

Pyramid B

- Large older population with approximately 25 per cent over 65
- Small percentage less than 15 years
- Relatively even distribution through youth and old age

b Pyramid A suggests a LEDC with a high birth rate and high but declining death rate. This accounts for the wide base and tapering apex. The high birth rate is typical of LEDC's with limited access to use of birth control. The relatively high death rate is a result of low standards of living and medical facilities.

The narrowing apex suggests lower life expectancy. The higher numbers in the 20–34 age group for males may be due to high immigration rates.

Pyramid B suggests a MEDC. The narrow base suggests a low birth rate, typical of MEDCs. This is due to better access to and use of birth control. Decisions by women not to have large families or any children. More women are career oriented and defer child bearing thus limiting their reproductive years. The pyramid bulges at the top suggesting a high life expectancy and an ageing population. More women are surviving to be 80+ years with social consequences. The widening at the top is the result of access to better health care and higher standard of living, and/or greater awareness of health issues.

c
- This country will have a high youthful dependency ratio.
- There is a high rate of unemployment among the young in a youthful population in developing countries.
- High levels of crime as a consequence.
- This will place a strain on the economically active population.
- The tax base will be small.
- The government may not have enough funds to provide schools, social services for the youthful population.
- This may lead to deficiencies in basic services including medical facilities, thus contributing to the high death rate, particularly high infant mortality rate, and continual high birth rate.

5 Population and Resources

General objective

On completion of this module you should understand the relationship between population growth and resources.

Specific objectives

You should be able to:

- assess the method that depicts population density
- explain the factors that influence population density
- explain the relationship between population and resources.

Introduction

This module examines the balance between populations and their available resources. The population density of an area is very important since it relates the number of people to the unit area on which they live. Density may be displayed on choropleth maps.

Some persons see the high level of population growth and consumption as outstripping production and leading to hunger (Malthus). Others more optimistically assert that human creative technology will allow them to overcome environmental limitations (Boserup). Ecologists apply the notion of carrying capacity to the size of the population that the resources of the environment can sustain.

Both positions have some merit: human populations need to live in sustainable ways in the face of finite resources while harnessing or developing new resources.

Content

- Population density and choropleth maps.
- Optimum population, under population and over population.
- Factors influencing changes in carrying capacity.
- Models of population growth in relation to resources – Malthus and Boserup.

Population Density and Choropleth Maps

Population density refers to the relationship between the size of the population and the area they occupy. It is defined as the number of persons per unit area and calculated by dividing the number of people in an area by its size in square units. Countries with large populations do not necessarily have high densities. Even small countries such as Barbados and Singapore, which have very high densities, show great internal variation in distribution (spread) since the calculation is an average.

At a regional level, urban areas have higher densities than rural areas. The vertical extension of urban buildings and the rural use of land for agricultural production account for these differences. Migration from rural to urban areas also influences the urban density.

Note: Density reflects intensity or concentration of people; while distribution (Module 1) is related to spatial location or arrangement. Densities are said be 'high' or 'low'; while distributions are 'even' or 'uneven'.

Refer to Figure 5.1 on page 36. The map on the left shows actual number of persons per ward. This map shows population distribution. However the one on the right shows persons/area, that is, density. Each map gives a different visual impression of the same population. Table 5.1 below gives population data for selected countries.

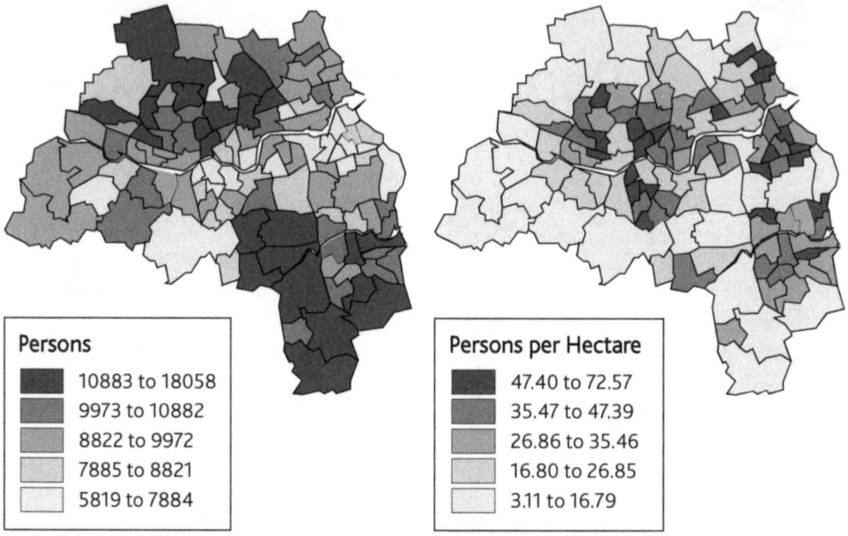

Source: Office for National Statistics

Figure 5.1 *Choropleth maps: A comparison of count data vs rate data: Tyne & Wear, Population by Ward, 2001*

Table 5.1 *Population data for selected countries*

Country	Population '000s	Area km^2	Density persons/km^2	Gross National Income Per capita US $
Haiti	8,900	27,750	321	480
Jamaica	2,700	10,901	248	3,480
Trinidad/ Tobago	1,300	5,130	253	13,340
Guyana	750	214,969	a)	1,130
Barbados	280	430	651	13,000
St. Lucia	120	616	b)	5,110
China	1,313,437	9,584,492	137	2,010
India	1,169,016	3,064,898	381	820
USA	305,826	9,826,635	c)	44,970
Japan	127,967	377,727	339	38,410
UK	60,769	243,609	d)	40,180
Singapore	4,436	639	6,942	29,320

Source: CIA World Factbook

Activity 5.1

1 *Distinguish* between population density and population distribution.

2 *Calculate* the missing population densities a) to d) on the Table 5.1.

3 Use evidence from the Table to support the suggestion that 'there is no relationship between population density and wealth'.

Feedback

1 Population density and distribution have some common features, for example both reflect spatial variation. However they can be distinguished by particular characteristics:

Density is related to the land area while population distribution is simple location of people within an area.

Population distribution is a visual attribute reflecting the spread; whereas density is amount/area or concentration.

2 a 3/km^2 b 195/km^2 c 31/km^2 d 249/km^2

3 It is clear from examples given in the Table that there is no relationship between density and wealth. The countries with the highest densities are neither the wealthiest nor the poorest. For example, Barbados with the highest density amongst the Caribbean countries, has one of the highest GNI per capita in the region, similarly Singapore with 6,942 persons/km^2 has a higher GNI than any of the Caribbean countries. In the Caribbean, Guyana has the lowest density and one of the lowest GNI in the region. The number of persons per unit area does not seem to predict wealth since it does not reflect the resources available in the area nor the ability of the population to use them.

Choropleth maps

Choropleth maps use intensity of shades of the same colour or line shading to reflect differences in density of population. This gives an immediate visual impression – the darkest shade represents the highest population density, lightest shade, the lowest density. The shade is uniform over each division within the given density. Ideally, no more than five shades should be used on any one map. The maps in Figure 5.1 are both choropleth maps, but numeric data are plotted on the map of the left, and density on the right.

Construction of choropleth maps

1 Collect data and calculate density/unit area for areas given (county/parish).
2 Choose a map of an appropriate scale with the boundary divisions identified but usually not named on the map.
3 Examine the extremes of data and note any breaks or clustering in the values.
4 Divide the data in groups, generally, not more than five with even intervals.
5 Pencil in the group on the map to check for visual effect.
6 Assign a shade of one colour to each group – darkest shade for highest intensity; lightest for lowest. Avoid black or white which gives the impression of all or nothing.
7 Check that map has a title, frame, key, scale and north arrow.

Merits and demerits of choropleth maps

Merits

- There is immediate visual impression of spatial variation of density by shade association.
- The actual density values can be identified.
- There is an objective calculation of values.

Demerits

- The sharp boundaries between divisions give a false impression of change.
- Equal densities over large areas hide within area variations.
- The image is dominated by large areas in the same shade.
- They show rates, that is, population/area rather than total values.

Note: it is desirable to use data from small areas if data are available. Line shading may be used instead of colour.

Activity 5.2

1 Define 'crude density' of population.

2 Using a well labelled diagram illustrate one method to depict the population density of a country using census data.

Feedback

1 Crude density of population is the number of people per unit area.

2

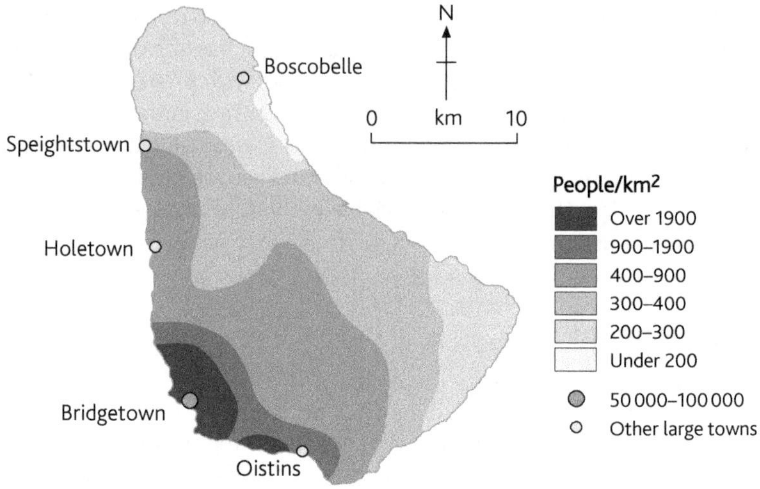

Figure 5.2 *Population density – Barbados*

Optimum Population, Under Population, Overpopulation

The ideal relationship between population and resources has been the focus of consideration for a long time. Generally some balance between population demand and resource supply is sought. But this simple picture becomes more dynamic when the distribution of population and consumption of resources are more closely examined. It is not simply a case of reducing the population so as to have universal availability of resources. A very small percentage of the world's population controls and consumes a high proportion of the world's resources. Concepts of optimum population; under population and over population are often applied to country comparisons, for example, 'Bangladesh is over populated and Canada is under populated'.

Optimum population

This assumes the ideal situation in which people use all resources to produce the highest standard of living (per capita economic return). Any change in either population or resource use would change the output and result in either over or under population.

Overpopulation

This should not be confused with either large populations or high population densities since some densely populated countries are able to use their resources to provide for a high standard of living. However, if the country is unable to meet the basic needs of the population, with widespread poverty, famine and a low standard of living, then it may be over populated, that is, there would be too many people for the resources of the area. Often there is accompanying environmental stress as, for example, in Haiti.

Under population

This concept suggests that an area can be considered under populated if more people using the resources could produce a higher standard of living.

Activity 5.3

Use the diagram below to answer the questions following.

1 Identify the points P, Q and R as one of (i) over population; (ii) under population; (iii) optimum population.

2 Explain how 'optimum population' relates to quality of life.

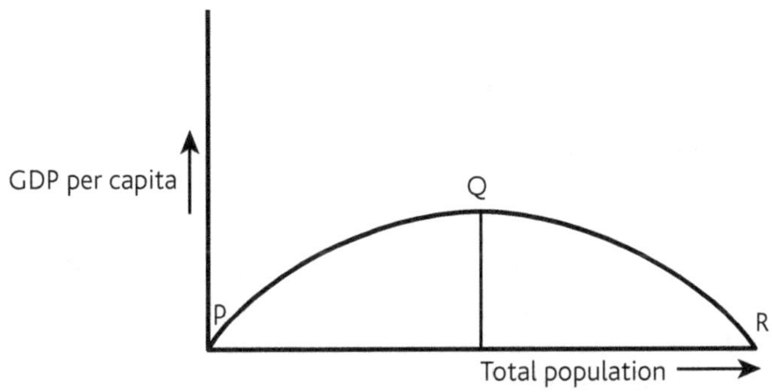

Figure 5.3 *Gross domestic product and total population*

Feedback

1 P – Under population

Q – Optimum population

R – Overpopulation

2 Optimum population relates to the quality of life because an optimum population is one that makes best use of the available resources, thus ensuring a higher quality of life for the population. Since resources are exploited to their fullest potential, there is neither wastage nor under-utilization of resources, resulting in an enhanced quality of life. A change in resources or population will result in over or under population.

Carrying Capacity

The ecological concept of carrying capacity is defined as 'the population of a given species that can be supported indefinitely in a defined habitat, without permanently damaging the ecosystem upon which it is dependent'. When applied to human populations it refers to maximum rates of resource use, waste production and sustainability without damaging the ecosystem. It also raises questions of what standard of living is to be created and for how many people. Human beings consume resources well beyond basic needs, but with creative technology they are not totally dependent on the environment as are animal and plant species. Human beings are the only species that can change the relationship negatively or positively between resources and populations, largely through the use of technology. The carrying capacity will vary depending on level of technology and capacity of population to manipulate the environment.

Factors influencing changes in carrying capacity

Any change in the population or resources in an environment will change the carrying capacity. These changes may be short term or long term.

- *Population size* – an increase in the population by natural increase or immigration will increase the demand for resources and affect the carrying capacity. Conversely if the population is decreased the demand would be lessened.
- *Environmental resistance* – changes in the environmental conditions such as food availability, natural disasters, diseases, climatic change will all negatively impact carrying capacity.
- *Resource innovation* – development and increased efficiency can all increase the carrying capacity of an environment.

Models of Population Growth in Relation to Resources: Malthus and Boserup

Malthus, 1798

In an *Essay on the Principle of Population*, Malthus suggested that the geometric increase in population would outstrip the arithmetic growth of food supply, and without the checks and balances of famine and war, a population would outrun the limits of the environment to support it. These checks could be positive, which referred to circumstances that led to a reduction in population size such as famine, wars and diseases. He also identified negative checks or those which limited growth such as late marriage and abstinence.

Boserup, 1965

Esther Boserup asserted that an increase in population would lead to increased innovation and invention. Necessity is the mother of invention. Her main arguments are that the higher population would be an incentive to technological development and technological development is a determinant of demographic change. Population change stimulates changes in agricultural systems leading, for example to agricultural intensification. It is the increased demand that is responsible for the pressure to change agricultural production. The growth of population and agricultural output has spin off effects which will stimulate the process of economic growth.

Malthus and Boserup approached the issue of population and resources from opposite ends. Malthus has been criticized because he did not foresee the technological developments that have been responsible for the large increases in food production today. He placed too much emphasis on reproduction as a cause of poverty and ignored the social and economic causes of poverty. Moreover, today, population growth is not regarded as a cause of development problems but rather as a symptom. The reason why the poor have large families may be rooted in the social and economic conditions of the countries and may be a logical reaction to those conditions.

Critics say that Boserup did not give credit to the fact that there are fragile environments that cannot support large populations. Her thesis may apply to agricultural societies in the developing world where the numbers that could be supported depend on agricultural technology. In modern industrial societies, inadequate food production may lead to out-migration rather than innovation. Some of her critics also argue that innovation does not occur immediately, and if population growth is rapid, adjustments may not have time to occur.

Malthus' theory, developed in pre-industrial society, still has some validity today. Unchecked consumption and runaway use of resources can lead to disaster as the demand for resources is greater than technology can supply. On the other hand Boserup's more optimistic approach has shown that in modern times human inventiveness has the ability to stave off or divert the disaster of scarce resources.

Activity 5.4

1 'Population growth leads to development'. State three main features of Boserup's view of the relationship between resources and population growth.

2 Outline one criticism of this theory.

Feedback

1 One feature of Boserup's view of the relationship between population and resources was that increased population pressure would lead to an increased demand for food which would force increased productivity and efficiency.

Secondly technological change is a determinant of demographic change.

Thirdly she asserted that the increase in population and agricultural output will have spinoff effects and these will stimulate economic growth.

2 One criticism of this theory is the severe imbalance and disparities in population growth and technology around the world. The poorest countries have the highest population growth rates but the lowest levels of technology hence famine still plagues areas of the world.

Key Points

- Population density is often shown using choropleth maps.
- Populations and resources are often unbalanced creating situations of under or over population.
- Many factors can change the balance in the carrying capacity of an environment.
- Some theorists are pessimistic about the future of human populations (Malthus) while others are more optimistic (Boserup).

Conclusion

It is a vital concern that human populations pay attention to the use of environmental resources. A balance between the demands of growing populations and finite resources requires careful consideration. Humans can overcome limitations because of their creativity, but still need to be mindful of the desirability of living in balance with their environments. Historically and spatially, a balance between population growth and resource use vary widely. The goal should be sustainable use of the world's resources.

End Test

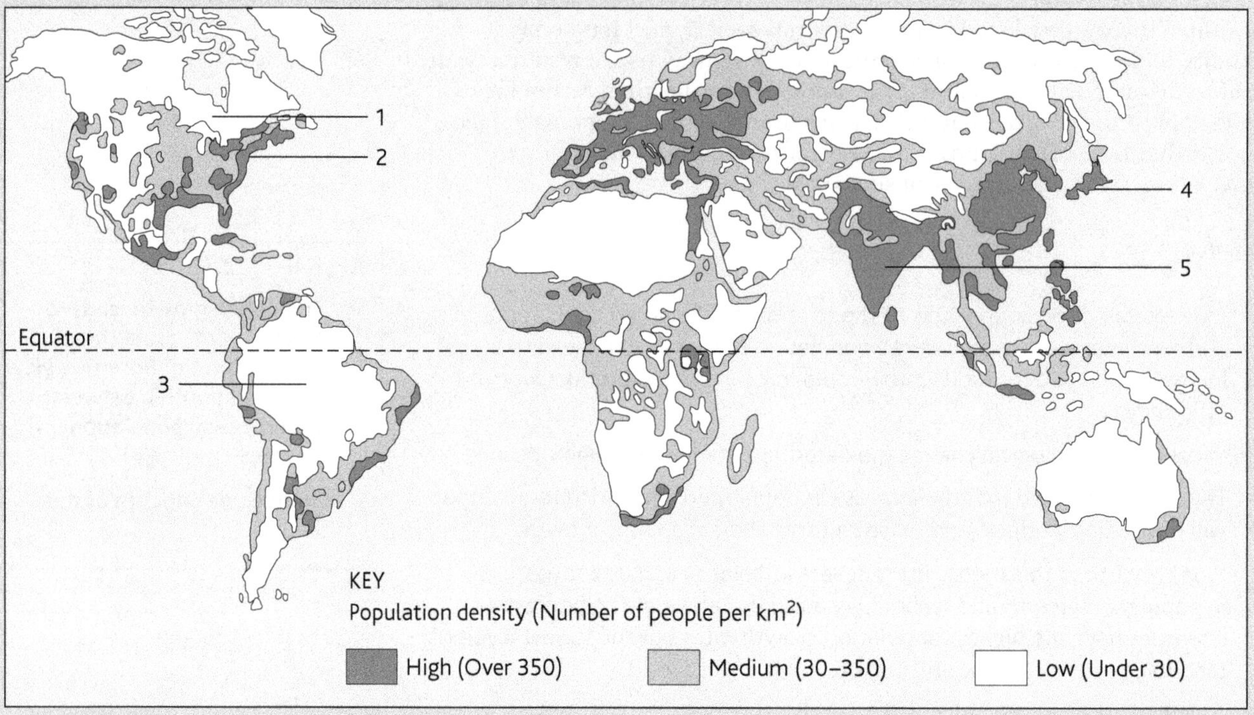

Figure 5.4 *Map showing distribution of population*

1 Name the areas numbered 1–5 on the map Fig 5.4.

2 Describe the variation in density of population shown in South America on the map shown.

3 Outline two reasons for the variation in the density of population in question 2. Maps of South America in your atlas should be able to help you to answer this question.

End Test Feedback

1 Areas numbered 1–5 on the map are:

   1 Northern Canada/Canadian Shield

   2 NE USA

   3 Amazon Basin/Central Brazil

   4 Honshu

   5 India/Deccan Plateau

2 Variation in density shown in South America: Greatest population is generally coastal. Concentrated in urban areas such as Lima, Sao Paulo and Caracas. Lowest densities are found in the rainforest and Southern Andes, with moderate densities on plateau areas like Brazilian Highlands and Bolivian Plateau.

3 Two factors which account for the variation in density of population in South America:

   Physical: the cold climate of the Andean region discourages settlement, so do the heavy rainfall and dense equatorial forests of the Amazon Basin. Much of the east coast, with milder climate and access to Atlantic trade routes, has higher population densities.

6 Settlement Processes

General objectives

On completion of this module you should:

- understand the factors affecting the forms and functions of human settlements
- develop appropriate skills and techniques in Human Geography.

Specific objectives

You should be able to:

- analyze the types, patterns and factors affecting the location of rural settlements
- analyze changes in rural settlements in MDCs
- apply urban models to the growth of the city and development of functional zones in LDCs
- explain the processes and problems of urbanization in MDCs and LDCs
- explain the solutions to urban growth in MDCs and LDCs
- use topographical maps to analyze settlement patterns.

Introduction

The change in human lifestyle from nomadic hunter/gatherers to sedentary cultivators was of great significance. The domestication and cultivation of plants and animals led to a food surplus so that some persons could do other tasks and specialize. The earliest settlements were located along rivers/springs and often near fertile alluvial/volcanic soils.

Settlements are generally dependent on some type of economic activity to support their populations: from agricultural-based rural settlements to the urban service-dominated skyscrapers. Settlements vary in size and function over a continuum from the single isolated dwelling to the largest city. They also form different spatial patterns.

There are models of urban growth which explore the development of functional zones in cities. Urban growth has resulted in both positive and negative effects.

Settlement processes have a great impact on the environment, changing and manipulating all aspects. Disposal of human waste products form a large problem for many cities and pollute air, water and land.

Content

Rural settlements

- Rural settlements-types and patterns.
- Factors affecting location of rural settlements.
- Changes in rural settlements in MDCs.
- Urbanization in LDCs and MDCs.
- Causes and consequences of sub-urbanization, counter urbanization, re-urbanization or gentrification in MDCs.
- Models of Burgess, Hoyt and Ullman and Harris and applicability to LDC cities.
- Solutions to urban growth in MDCs and LDCs.

Rural Settlements

There are many differences in the way in which countries define urban and rural. In general, the criteria for distinguishing rural from urban are population size and density and the availability of certain services. However the population thresholds differ. In some African countries, the population threshold for an urban settlement is 100 persons. In others, it is 20,000. In some countries in Europe, it varies between 2,000 and 2,500. In Jamaica it is 2,000. In Britain some cities are simply defined as towns with city status.

The question as to what makes an area rural is even more difficult. In most rural areas, primary activities – agriculture, forestry, mining – occupy extensive areas of the land. Settlements derive most support from these primary activities, the products of which are processed in urban areas. Because these activities take up a large proportion of the land in rural areas, rural settlements are more diffuse. However, improvements in transportation and communication are changing the nature of rural areas and the relationship between land and people.

Rural settlement types

Most studies of rural settlement types were devised in the context of specific regions – Europe, North America, China – and attempts to produce a general typology are very complex. However, most classifications are based on the relative location of individual buildings and there is a basic distinction between nucleated and dispersed settlements. In between these extremes there are semi compact and hamleted settlements.

Nucleated settlements are compact and the built up area close. In Jamaica, nucleated rural settlements were formed by freed slaves who were either given access to, or settled on land in close proximity to the plantations where some continued to work. In Trinidad and Tobago nucleated rural settlements were also formed by East Indians who, on completion of indentured service, were either granted or allowed to purchase blocks of Crown Land in lieu of return passage to India. They established their own villages.

In semi-compact settlements, there is fragmentation of nucleated settlements.

In hamleted settlements there is fragmentation into several units with basic services only. In India, these units are often based on caste.

Dispersed settlements are characterized by isolated households or farms and are found typically, in rugged upland areas. Freed Africans dispersed themselves in the uplands of Jamaica and in the interior of Guyana where runaway slaves had also settled.

These formed a settlement hierarchy with the individual household/farm at the base and the nucleated village at the top. Market towns are sometimes classified as rural.

Form/pattern

In nucleated rural settlements buildings are clustered around a feature. They may be:

- Linear, following roads or rivers.
- Squared, built around a square or green.
- Star-shaped and cruciform, built around a convergence of roadways.

Settlements may be dispersed:

- linearly
- randomly or
- clustered into hamlets.

Location/situation

The site of a settlement refers to the actual land on which the settlement was built. Its situation is its position relative to the surroundings.

Early settlers took into consideration the physical conditions as well as the economic potentials of an area for settlements. Early Spanish settlers in the Caribbean were sometimes guided by the conditions set out in the

Laws of the Indies. These laws, which were passed in Spain in 1573, were designed to regulate the siting and orientation of settlements to ensure uniformity in the Spanish colonies and to avoid conflict with indigenous populations. Some of these guidelines were followed in the selection of the site for the settlement of Spanish Town (Villa de la Vega) in Jamaica and St Joseph (San Jose' de Oruña) in Trinidad.

- Protection from the English and Dutch was important. Therefore, if possible, sites in 'maritime locations' were to be avoided. The early settlements at Spanish Town in Jamaica and St Joseph in Trinidad had inland locations. Protection was also needed from the prevailing winds and settlements were located on the leeward coasts of the islands where there was shelter from the trade winds.
- Sites that are too high and which may be affected by winds, or too low, since such sites are usually unhealthy, should be avoided. Choose medium elevations that enjoy 'good winds'.
- Proximity to fertile soil and land for farming and pasture.
- Fresh water for drinking and irrigation.
- Forests for wood and building materials.
- Ease of transport.
- A healthy location which could be judged from the abundance of old and young men of good complexion. Wetlands which harbour diseases should be avoided.
- There should be good outlet by sea and land.
- There must be people to work and cultivate the land.

The location of St. Joseph in Trinidad was described as being:

> 'abundant in provisions and there are many lands which can be sown, pastures, water and common lands and other things necessary to found a settlement.'

The following was the description of Spanish Town, Jamaica:

> 'The land is plentiful in bread and beef and is healthy and that all who reside there have a healthy and easy life. It is a land of very good water, without mountains or ranges of hills and has very good ports suitable for navigation for the provinces of Santa Marta and Cartagena.'

It is clear that the location of these two settlements was influenced by the Laws.

Under British rule, trade with the Mother Country was extremely important and settlements were developed in coastal locations. The settlement at Bridgetown in Barbados was founded at an Amerindian bridging point over an arm of the sea. An extensive colonial network of settlements was established by the British through which trade was organized. These ultimately became the port cities of Bombay, Calcutta, Singapore, Lagos, Rangoon, Kingston and Bridgetown.

The Laws of the Indies covered a host of physical, human and economic factors – climate, soil, vegetation, elevation, access to water, transportation, disease, labour supply. These are the factors that influenced the location of early settlements in all countries. The selection of sites in Europe was influenced by the opportunities for trade presented by bridging points, the confluence of tributaries. Settlers selected positions which they could easily defend; wet points for access to water, dry points to escape flooding and land that was suitable for cultivation.

Function

Most modern settlements, even those in rural areas, are multifunctional and are classified by the dominant function.

1 Market/agricultural centre
2 Dormitory
3 Route centre
4 Port
5 Tourist centre

The line between traditional definitions of rural and urban is blurred by developments such as 'suburbanized villages' in MDCs and tourism centres in LDCs.

Changes in Rural Settlement in England

Some rural areas in England are growing more rapidly than urban districts and this growth has been fuelled by internal migration. The pattern of rural growth varied. Rural areas accounted for about 19 per cent of the population in 2005 and while such areas, as a whole, have been growing, about 61 per cent of this growth has been experienced in accessible rural areas. In addition, there are areas within remote rural areas where the population is declining.

The counter urbanization trend, discussed later in this module is responsible for the increase in the population of rural and the decrease in large metropolitan areas in England. Between 1981 and 2001 Metropolitan England lost 2.25 million persons to internal migration and this movement has been large enough to influence the demographic character of rural Britain. There has been a strong counter urbanization trend among those in the over 60 age group and ageing rates are higher in rural than urban areas. On the other hand, rural areas have been losing those in the age group 16 to 24. The median age in urban areas in 2001 was 38.5 but 44.4 in rural areas. However, the contribution of migrants in the over 60 age group to rural growth was less than those in the 30 to 44 and 45 to 59 age group. Retirement, therefore, was not the major factor accounting for rural growth and net in-migration into accessible rural areas is often accompanied by net out-commuting into metropolitan areas for work.

Migration to rural areas is not only age selective. It is also racially and socially selective. The urban exodus is predominantly United Kingdom-born and white. So the urban rural differential for race and ethnicity is also steepening.

Moreover, the competition for scarce housing in rural areas is driving social change. Because of housing costs, the more affluent is replacing less affluent households. It is usual to think of gentrification (discussed later) as an urban process. However, a similar process – rural gentrification – is occurring in some parts of England. Many leave the cities to escape the diminishing quality of urban life and are drawn to what they perceive as 'green' residential space. Some establish primary but also second homes. They renovate old farm buildings and transform the built environment as well as the 'natural'. They create 'manicured' spaces. They introduce new species of flora and fauna, many of which escape from gardens into the surroundings. Favoured species are encouraged by fertilizers, and weeds and pests treated with herbicides and pesticides. They simplify and eventually obliterate much that was 'natural'.

The countryside may have been built on agriculture and forestry but this foundation has been eroded in England by new technologies, increased affluence, increased mobility and the desire for leisure. While differences in age and ethnicity between urban and rural areas are growing, urban and rural economies are converging. About 73 per cent of jobs in rural areas are now in the service sector. Information and knowledge-based industries are playing an increasing role. In addition, incentives have been given to farm diversification and this has taken the form of retailing in farm shops and craft centres, food processing such as the making of jams, the provision of accommodation such as bed and breakfast, sports, recreation, farm museums. Rural areas exploit biodiversity, scenery and cultural heritage. When the outbreak of foot and mouth disease in Scotland in 2001 virtually closed rural areas, loss to tourism was estimated at between 200 and 250 million pounds sterling.

But not all rural areas are benefiting from these developments. There are remote rural areas experiencing population losses from net out-migration and natural decrease. Birth rates are falling and death rates among elderly populations rising. There has been a loss in farm employment with increasing mechanization and the pull of the attractions of urban areas. Settlements beyond commuting range are not attractive to those in the working age group. In the residual rural population, the old and the very young are over represented. There is a housing problem caused by competition from in-migrants who are fragmenting the local community. As the population falls, there is a decline in service provision – public transportation, post offices, general stores, schools and health centres. The decline of these services creates problems for an ageing population, many with special needs. These problems are sometimes difficult to address because deprived areas may not be concentrated but dispersed throughout rural areas.

One approach to these problems is the key settlement policy in which resources are concentrated into selected rural centres to stem or reverse rural depopulation. The objective was to limit migration to internal movement from very remote rural areas to the key centres.

Activity 6.1

For any MDC investigate changes in its rural settlement by doing the following:

1. Collect data on the size of villages in a particular administrative area over the last 20 years.
2. Calculate the percentage change in population over the area and identify the trend of depopulation or repopulation.
3. Locate the settlements on a map and measure the distance of each one from the nearest large town. What is the relationship between the trend in change of population size and distance from nearest urban area?
4. Write an essay on the changes in rural settlement and reasons for them based on your research.
5. Research: How successful was the key settlement policy in England and Wales?

Feedback

Depending on area chosen, there should have been an inverse relationship between distance from nearest urban centre and growth. Generally rural settlements nearer to urban areas have been overtaken by urban sprawl and suburbanization. Those further away have been more subject to depopulation and so have declining populations.

The investigation of economic activity, would likely show increases in tertiary and quaternary economic activity at the expense of agriculture. There may also be changes in transport and communication making some rural settlements more accessible to a mobile population.

The Process and Problems of Urbanization in MDCs and LDCs

Urbanization is an increase in the proportion of the population that lives in towns and cities.

Table 6.1 *Growth of urban population by development region (billion)*

	1950	1975	2007	2025	2050
World	0.74	1.52	3.29	4.58	6.40
More developed	0.43	0.70	0.91	0.99	1.07
Less developed	0.31	0.82	2.38	3.59	5.33

Source: The World Urbanization Prospects: The 2007 Revision, United Nations

Table 6.1 shows the growth of the world's urban population between 1950 and 2007 and projections for 2025 and 2050. Between 1950 and 2007, the urban population of the world more than quadrupled and the annual rate of increase averaged 2.6 per cent. By 2050, the world's urban population is expected to be the same size as the total population of the world in 2004.

In 1975 there were more people living in urban areas in the LDCs than in MDCs and it is estimated that there will be five times more people living in towns and cities in less developed countries in 2050. Between 2007 and 2050, the rate of growth is expected to be about 1.8 per annum and the urban population is expected to double by 2050. The rate of urbanization is therefore expected to slow down. The reduction in urban growth is expected in both MDCs and LDCs (Table 6.2).

Table 6.2 *Rate of urbanization by development region (%)*

	1950–1975	1975–2007	2007–2025	2025–2050
World	0.99	0.88	0.82	0.59
More developed	0.97	0.33	0.33	0.24
Less developed	1.62	1.51	1.08	0.78

Source: The World Urbanization Prospects: The 2007 Revision, United Nations

Although the number of people living in towns and cities in less developed countries is greater than that in the more developed, the percentage is less. In 2007 the percentages for MDCs and LDCs were 74 and 44 respectively (Table 6.3). Although the disparity is expected to be less, the more developed countries are expected to have a higher proportion of its population living in urban areas in 2050. MDCs are therefore more urbanized than LDCs.

Table 6.3 *Percentage urban population by development region*

	1950	1975	2007	2025	2050
World	29.1	37.3	49.4	57.2	69.6
More developed	52.5	67.0	74.4	79.0	86.0
Less developed	18.0	27.0	43.8	53.2	67.0

Source: The World Urbanization Prospects: The 2007 Revision, United Nations

However, levels of urbanization are not the same throughout the developing world. Latin America and the Caribbean had a very high level of urbanization – about 78.3 per cent – in 2007. In other words, the level of urbanization in this area of the world is higher than in Europe (72.2 per cent) and Oceania (70.5).

Africa and Asia, with 38 and 41 per cent of their population living in urban areas are still largely rural.

Megacities

In 2007, 19 cities were regarded as megacities because they had a population of at least 10 million persons. In 1950 there were two (New York and Tokyo) and in 1975, Mexico City joined this group.

Eleven of the world's megacities in 2007 were in Asia and only one in Europe. Table 6.4 shows that Tokyo is expected to be the largest city in 2025 followed by cities in India and Bangladesh. These megacities only accounted for 9 per cent of the world's urban population in 2007 and the percentage is estimated to be about 10 in 2025. Most of the world's urban population live in cities with fewer than a half a million inhabitants.

Table 6.4 *World's six largest megacities, 2007 and 2025 (population in millions)*

2007		2025	
Tokyo	35.7	Tokyo	36.4
New York	19.0	Mumbai	26.4
Mexico City	19.0	Delhi	22.5
Mumbai	19.0	Dhaka	22.0
São Paulo	18.8	São Paulo	21.4
Delhi	15.9	Mexico City	21.0

Source: The World Urbanization Prospects: The 2007 Revision, United Nations

In MDCs, a longer period of urbanization, transportation changes and technology have allowed the decentralization processes of suburbanization and counter-urbanization to reverse the historical trends of moving to large urban areas. The movement represents a shift in population down the hierarchy. In both processes people move out of the cities.

- Suburbanization is the movement of people out of the inner areas of the city to the edges. Growth of fringe areas is more rapid than in the core. Increasing affluence, improvements in transportation, the development of housing estates and a desire to escape the deteriorating neighbourhoods in the city, have all fuelled this movement. It results in the development of suburbs, an extension of the built-up areas and a lifestyle involving daily commuting to work in the city.

  In some areas, suburbanization involves two processes. Suburban intensification or the infilling of suburban areas. Suburban extensification is producing the urban sprawl. The two processes are responsible for the development of what has been described as edge cities, attracting offices, shopping malls and recreational facilities.
- Counter-urbanization is, as the name suggests, the movement in the opposite direction to urbanization. People and businesses move into rural areas, to some extent, changing the nature of the village to an outlying 'suburb' beyond the suburbs. It is facilitated by decentralized workplaces and high technology, for example, footloose computer jobs. It also represents a desire to be closer to nature.

At the same time, there is re-urbanization, a process designed to improve the quality of inner city life for people and households in different social strata. It is an attempt to provide vibrant living space within urban cores that have fallen into decline. Gentrification is a form of this process. Previously abandoned buildings are re-furbished and specialized commercial services encouraged until the area becomes a desirable location for the wealthy. It involves refurbishment and change in social composition through the displacement of the original inhabitants and immigration of middle class households. As a result, many city centres have become socially exclusive.

Problems of urbanization

LDCs

Early urbanization in more developed countries resulted from industrialization and the creation of new job opportunities. Migrants provided labour for the new factories. In less developed countries today, urban growth is taking place in the absence of significant industrial development. Job opportunities are, however, concentrated in urban areas and rural areas are in decline. Migrants contribute to the problem of surplus labour in cities in less developed countries creating low wages for those employed and high levels of unemployment. Many work in the unregulated informal sector where wages are low and where there are no social security benefits.

The rapid growth of population in towns and cities in LDCs is responsible for many environmental and social problems. Countries lack the capacity to provide services.

- *Housing*. In many urban areas in developing countries shanty towns and squatter settlements occupy the urban periphery. Housing is of very low quality usually self-help, using material at hand, for example, the favelas of Brazil. Homelessness may be very high as in Calcutta where many thousands of people live on the pavements.
- *Lack of clean water*. Less than 5 per cent of population of Lagos (Nigeria) has access to piped water. In Delhi water is available for only part of each day to the 50 per cent of persons connected to public service. Childhood diseases such as gastroenteritis are problems in such environments.
- *Sanitation*. Poor sanitary conditions in many of the unplanned squatter settlements. Raw sewage and garbage flow in open trenches.
- *Transportation*. The congestion is acute as large numbers of persons move on foot and on overcrowded public transport. Long traffic delays result.
- *Social services*. Urban poverty is reflected in lack of educational and healthcare facilities. Child labour and disease resulting in high death rates in some shanty towns.
- *Air pollution*. The exhaust of motor vehicles in cities such as Bangkok, in Thailand, produce a cocktail of gases that is responsible for high levels of respiratory complaints. Vehicles are poorly maintained as few countries have emission standards. Particulate air pollution is extremely high. In Rio de Janeiro, it is twice as high as in New York and four times as high as in Paris. Respiratory illnesses are a problem.
- Rapid urbanization is also creating social inequalities that are associated with crime.
- Urban sprawl is a problem in many countries. In Brazil it is placing pressure on the natural habitat and much of the rainforest is being lost. With it, there is a loss of biodiversity.

MDCs

In MDCs, urbanization has resulted in homelessness and inner city slums. Problems in MDCs include:

- *Housing*. Old and poor quality in the inner cities (adjacent to the CBD) with some dereliction and abandoned buildings. Unsightly graffiti may mar the walls. Ghettos may form in cities with high migrant populations. Overcrowding and high densities occur in high rise buildings after slum clearance.

- *Lack of basic amenities*. Many old buildings are in a poor state of repair.
- *Recreation*. Lack of recreation areas and spaces.
- *Pollution*. Air quality is low and there is environmental pollution. Air and noise pollution is common in the heavy traffic congestion of the inner city.
- *Economic*. Decline in manufacturing industries has lead to high unemployment. New industries have been established on the edges of the city where land is available at lower cost. The problem is worsened by new globalizing trends in industry such as outsourcing (contracting to a supplier) and offshoring (the transfer of functions to another country). Many labour intensive activities are outsourced to low wage countries and it is believed that this has increased unemployment among the unskilled.
- *Social*. There is a high concentration of ethnic minorities and low income families in inner city areas where unemployment is high. Crime is a problem among the youth gangs that control many areas within the inner cities. Drug use is also a problem.
- *Urban redevelopment* in some cities has resulted in the exclusion of the original inhabitants who cannot afford the cost of housing in these areas, for example, London's Docklands. Many close knit communities have been destroyed to make way for the new industries and luxury accommodation.
- *Urban sprawl*. Cities such as Los Angeles face the problem of urban sprawl, a pattern of land use, transportation and social and economic development. Trips, even for low order goods, require the use of a vehicle. The urban heat inland is extended. There are threats to air quality and quantity and long commuting affects mental health.

Solutions to the Problems of Urban Growth in MDCs and LDCs

It should be noted that urban growth includes growth of urban populations by natural increase and migration. In both MDCs and LDCs similar solutions have been tried with varying success. These include:

- Construction of new towns.
- Decentralization of services and economic activity.
- Demolition of areas of urban decay.
- Improving facilities in rural areas.
- Redevelopment of some urban areas.
- Self help initiatives – for example, to improve housing.
- Limiting traffic into cities by road pricing, road space rationing, park and ride facilities.

City Size and Hierarchies

Zipf's Rank Size Rule

In all countries, small cities are more numerous than large. In countries with a long urban history, the size hierarchy is said to follow the rank-size rule.

Zipf's Rank Size Rule examines the relationship between the sizes of towns in a country. In 1949, he suggested that if the towns of a country were ranked in descending order of population size, the size of population will decrease in inverse proportion to the size of the largest divided by its rank. For example, if the largest city has 10 million people then the 2nd ranked would have 10 million divided by 2 = 5 million; 3rd ranked would have 10 million divided by 3 = 3.3 million and so on. This relationship can be shown on semi-logarithmic paper, since the relationship between population sizes is not arithmetical.

Urban primacy

In developing countries, Zipf's Rank Size Rule hardly applies, as a combination of socio-cultural and historical factors results in the dominance of one very large urban area. It is often the administrative centre and main port that dominates commercial activity of the entire country. This city is referred to as the primate city and is more than twice as large as the next largest town. Most Caribbean countries are dominated by one very large town. Primacy also occurs in developed countries, for example, London which is six times larger than the next largest city in England; and Paris which is ten times larger than the next largest city in France.

Table 6.5 showing the largest towns in Jamaica, illustrates Jefferson (1939) Law of the Primate City. Kingston is nearly four times as large as the next largest town, and all the other towns are smaller than Zipf's rule would predict. In fact the spatial and economic primacy becomes even more severe if the location of the largest three cities is considered. They, in fact, form a continuous urban/industrial corridor to the west of Kingston, therefore fully one third of the country's population live in this conurbation. This concentration of population in a few urban centres is typical of developing countries, where continued disparities exist between rural and urban areas. Jamaica is just 57 per cent urbanized and migration from declining rural areas to urban areas would continue to exaggerate this primacy.

Table 6.5 *Population of largest cities of Jamaica*

City	Pop (000s)	Rank	Zipf size
Kingston	578	1	578
Portmore	160	2	489
Spanish Town	131	3	193
Montego Bay	96	4	144
May Pen	57	5	113
Mandeville	48	6	96
Old Harbour	24	7	82
Savannah la Mar	20	8	72
Ocho Rios	16	9	64
Port Antonio	15	10	57
Linstead	15	11	53
Santa Cruz	11	12	48

Source: National Census Report 2001, Jamaica

Urban Models: Burgess, Hoyt, Ullman/Harris

Urban models or models of urban morphology, investigate the spatial arrangement of functional zones (activities) in an urban area. They assume spatial specialization where similar activities will locate in similar places. The three models discussed here are early attempts to understand sociological patterns in US cities: Burgess 1924, Hoyt 1939, and Ullman/Harris 1945, who worked mainly on North American cities.

The main functional zones identified in the models are:

- CBD
- light manufacturing
- heavy manufacturing
- lower income housing
- middle income housing
- high income housing.

Burgess (1924) – concentric rings

Burgess suggested that cities develop in concentric rings outwards with decreasing intensity of land use. His central area, the CBD, was the concentration of commercial activity. Around it were light manufacturing and the low income housing of workers who needed to live close to their factory jobs. Further away was the more affluent middle and upper income housing. Heavy manufacturing and the start of the suburbs were found on the outskirts. Migrants moved into the lower income housing and the city grew outwards.

Hoyt (1939) – sector

Like Burgess, Hoyt also had the CBD at the centre of his model and retained concentric semi-circles of light manufacturing and middle income housing around the CBD. However, he proposed wedges of growth associated with major roads. His areas of heavy manufacturing and low income housing formed wedges extending outwards from the CBD. The upper income housing also was located wedge-shaped along another road away from the manufacturing areas. The process of urban land use was controlled by the location of the main routeways.

Ullman/Harris (1945) – multiple nuclei

Ullman and Harris, retained the CBD but suggested that it was not the only centre of activity. They proposed other centres or multi-nuclei around the city. Older centres, industrial estates and shopping centres may each have their own specialized urban functions around them. The resultant spatial arrangement is more a patchwork of different uses. However, like the other models, lower income housing remained nearest light manufacturing with middle and higher income further away from the CBD.

Activity 6.2

1 Urban morphology refers to a city's:

a suburbs b growth c form d size.

Refer to the urban model diagram below to answer the following questions.

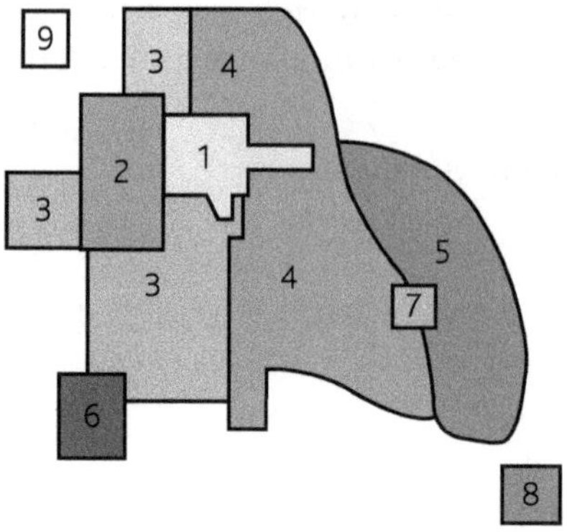

2 The zone numbered 3 is the:

a CBD
b light manufacturing
c low-class residential
d medium-class residential.

3 This model was developed by:

a Burgess b Hoyt c Mann d Ullman/Harris.

Investigate: the causes and consequences of gentrification in a specific city.

Feedback

1 c 2 c 3 d

The Applicability of Burgess, Hoyt and Ullman/Harris to Cities in Developing Countries

There are some basic problems with applying these models to urban land use in developing countries:

- Urban growth has been much more recent and rapid in developing countries than in developed countries of the previous centuries.
- Economic development and diversification has lagged far behind the demands of urbanization in developing countries.
- Colonial history created distinct 'new' and 'old' sectors of many cities with almost parallel and distinct economies and land use.
- Assumptions about availability of individual and/or mass transport are often not applicable to cities in the developing countries.
- Greater government/planning controls on urban land use in developing countries than in developed countries in the past.
- Many cities in developing countries have large illegal squatter settlements in extreme poverty.
- Many areas of cities in developing countries are multifunctional with the lack of specialization proposed by the models.

Key Points

- In all countries, the location of settlements is influenced by physical, human and economic factors.
- In MDCs, rural and urban economies are converging.
- By 2050, the world's urban population is expected to be the same size as the total population in 2004.

- Models of urban morphology help in the investigation of urban processes.
- Suburbanization and counter urbanization are reversing the historical trend of movement into cities.

Conclusion

An understanding of settlement processes is vital to the survival of human populations. As the world becomes more and more urbanized it becomes critical to seek solutions to the problems created by such high densities of population. The process of urbanization creates many environment stresses and urban deprivation which are threats to the sustainability of cities.

End Test

1 Explain how four factors contribute to the increasing importance of counter urbanization in developed countries.

2 a Identify two housing problems that are common in cities in the developing world and the developed world.

b Compare two housing problems of a named city in the developing world with those in a named city in the developed world.

c Compare one response of each city named above in dealing with the housing problems.

End Test Feedback

1 Counter-urbanization is becoming increasingly important in developed countries for a number of reasons:

Technology

- improvements in technology such as computer technology which means that people can work from home (not restricted to the city) by using computer technology;
- improved communication network.

Transportation

- improvements in public transportation – cars and trains, people can live in other areas and can still quickly and easily maintain contact with place of work in the city;
- greater personal means of transportation.

Push factors operating in the city – overcrowding, traffic problems, pollution, urban blight, high crime rates, restricted land area for expansion, housing available, high cost of housing and commercial space.

Shift from manufacturing to services and better wages in services.

Services not tied to original location factors but are more foot-loose.

Pull factors operating in the smaller settlements – less crowded conditions, more pleasant/greener surroundings, larger land area for expansion, lower cost of housing and commercial space.

Less traffic congestion.

Government policies such as those relating to the construction of new towns and new cities in Britain.

2 a

Developed	Developing
Location – Inner city slums. – Single residence converted to multiple use.	– Inner city slums. – But also squatter settlements on the outskirts of city or unused/uninhabited land or gully banks, unstable slopes. – Often these locations are hazardous.
Quality – Buildings several stories high, 'walk-ups' posing problems for the elderly. – Do not meet health/building standards. – Many are damp, rat-infested.	– Poor found in similar conditions in inner city tenements. – In KMA, they are also found in urban yards. – Large numbers of households share rudimentary toilet/sanitary facilities. They lack access to running water. – Waste disposal poses a problem.

2 b

Developed (London)	Developing (Kingston)
Building materials – Usually solidly built but in poor repair. – Faulty electrical connections cause fires.	– Do not meet existing building codes. – Built of whatever material is at hand. – Often a fire hazard and fire is a constant threat.
Home ownership – Britain sold council houses, many in poor condition, to those living in them and they were encouraged to upgrade.	– Squatter upgrading programmes in which services provided. – Homeowners encouraged to upgrade homes. – Provision of sites and services and own building encouraged.

2 c

Developed	Developing
Urban renewal – Was put in place in several cities such as London. – However many renewal schemes were not geared to the needs of the poor. – They were taken over by the young upwardly mobile and this resulted in conflicts as in East London.	– Urban renewal in Kingston was attempted but with limited success. – It was restricted to very small areas surrounded by slums. – High levels of crime prevailed.

7 Hydrological Processes

General objectives

On completion of this module you should:

- develop an understanding of hydrological processes
- develop appropriate skills and techniques in geography.

Specific objectives

You should be able to:

- explain the main concepts, flows and processes associated with the hydrological cycle
- explain hydrological processes which influence the development of related landforms
- analyze the physical, biotic and geological factors which affect processes operating within the drainage basin
- explain the effects of human and physical interaction within the drainage basin over time
- apply concepts and processes related to the drainage basin at different geographical scales
- apply map reading skills, appropriate investigative and practical techniques.

Introduction

Hydrology is the scientific study of water. Water is important not only to human activities but also a very important agent in shaping the landscape in many environments. In the Caribbean, some countries are designated water scarce while others experience a surplus. Some Caribbean countries have complex surface drainage systems, for example, Guyana, Jamaica and Trinidad and Tobago; while others, such as Barbados, have underground drainage systems. This section looks at the paths which water takes particularly on reaching the earth's surface, but also as it moves through the air and sea in the hydrological cycle.

Content

- The hydrological cycle and the river basin.
- Factors influencing the hydrological cycle.
- Storm hydrograph and water budgets.
- Climatic, physical and biotic (human and vegetation) factors affecting drainage basin characteristics and flows.
- Factors influencing drainage patterns and drainage density.
- Use of topographical maps to identify geological, physical and biotic influences on the drainage basin.

The Hydrological Cycle

The hydrological cycle is a model which describes the movement of water through the atmosphere, cryosphere, hydrosphere and the earth's surface. It is a closed system in which water circulates continuously without gains or losses. The amount of water is fixed.

Water states

Water is found in different states throughout the cycle.

Water evaporates from the oceans and water bodies to form water vapour (invisible gas).

Water vapour condenses to produce water droplets (clouds, rain).

Water droplets may freeze to produce ice (snow, hail).

Water storage

Water is stored mainly in the oceans and seas, as well as ice sheets and glaciers. Surface and underground water (ground water) storage account for a small proportion of the total water storage. These are the reservoirs. Table 7.1 shows the relative size and residence time of the major reservoirs. Oceans are by far the largest and the biosphere the smallest reservoir. The residence time is the average length of time that a water molecule spends in a reservoir before visits to others. For example, the water in lakes is renewed approximately once in every 50 to 100 years. Some deep aquifers contain fossil water. Water that is evaporated quickly falls as precipitation so the average atmospheric residence time is just about 9 to 10 days. Thus, the different sources of water in the hydrological cycle are replenished at different rates. These are averages and the actual residence time may, under certain circumstances, be very far from the average.

Table 7.1 *Volume and residence time of major reservoirs*

	Volume (10^6km^3)	Percentage of Total	Residence Time
Oceans	1350	97.3	3,250 years
Glaciers	29	2.1	10^3–10^5 years
Aquifers	8	0.6	2 weeks–10^4 years
Lakes	0.1	0.01	50–100 years
Soil moisture	0.1	0.01	52 days
Atmosphere	0.013	0.001	8–9 days
Rivers	0.002	0.0002	2–6 months
Biosphere	0.001	0.0001	6 days

Source: Adapted from www.chemgapedia.de

Water flows

Water and energy are transferred from one reservoir to another by means of evapotranspiration, condensation, precipitation, infiltration, percolation and runoff. These are flows.

Evapotranspiration

This is the conversion of water to gas. About 80 per cent is from the ocean and it is at its maximum under conditions of high temperature and strong winds. Evapotranspiration is the net effect of evaporation and transpiration (water loss through pores). Actual evapotranspiration increases as temperature increases provided that the water is available. In other words, actual transpiration depends on the field capacity of the soil. Evapotranspiration also varies with the type of vegetation and the length of the growing season. Potential evapotranspiration is the amount that would be evaporated if the water supply were unlimited.

Condensation is the reverse of evaporation, the change from a gaseous to liquid phase. It occurs when warm air rises and cools as a result of convection, convergence, frontal and orographic uplift.

Precipitation may take the form of rain, snow, sleet, and hail with the type depending on the geographic location and the season. It provides the moisture which flows into streams and rivers and infiltrates into the ground.

Infiltration is the process by which water enters the ground surface. As water nears the surface, it is intercepted by plants, the amount depending on the type of vegetation. The water reaches the ground as stemflow and enters or infiltrates the soil. When the infiltration capacity is exceeded the water flows over the ground as overland flow. The water which infiltrates the surface supports plants, supplies water to wells, streams and springs. The maximum rate at which water can enter the soil is called the infiltration capacity (mm/hour). The rate of infiltration depends on the nature of the soil surface (whether it is bare or vegetated) and the type of soil. Fine textured soils, with high clay content, will quickly become saturated and impermeable, allowing maximum run off. Coarse soils and well jointed rocks such as limestone would be so permeable and allow very little water to remain on the surface. The type of vegetation and antecedent moisture in the soil are also important controls.

Percolation is the downward movement of water through the soil. Once in the soil, the water may either flow laterally as through flow or percolate deeper where it reaches cracks fractures and fissures where groundwater is stored. The upper surface of the groundwater store is the water table. Rocks which store water are called aquifers; while those which block water are aqualudes. Some geological series may create artesian basins where water in wells drilled will rise to the surface under its own pressure. These are called artesian wells and water bearing geological structures are very important sources of water.

The underground areas can be divided into vadose or non-saturated zone and phreatic or saturated zone. Water may then flow laterally to streams and rivers as groundwater or base flow.

Runoff is the movement of water across the earth's surface towards rivers, lakes and oceans.

Figure 7.1 is a diagrammatic representation of the hydrological cycle.

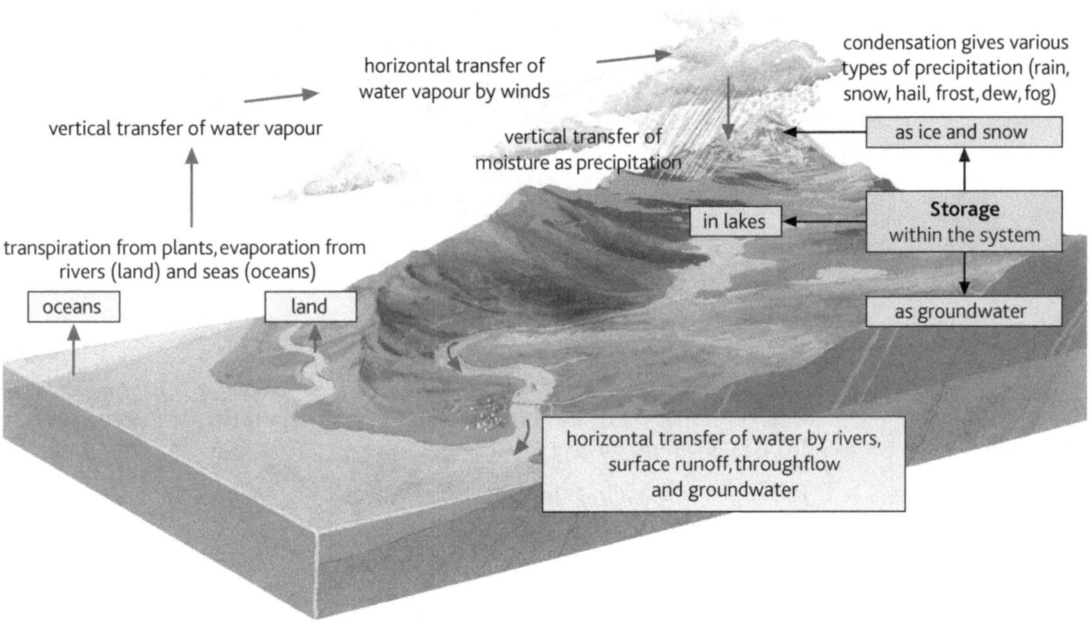

Figure 7.1 *The hydrological cycle*

Human activity

Water is essential for human life and domestic purposes. Reservoirs and dams greatly impact on the water cycle. It is also used for many other human activities such as agriculture, manufacturing and even tourism/leisure activities.

Water is pumped from wells which draw upon underground sources; is stored in reservoirs; channelled in drains; used in households and industries and released (with/without pollutants) into rivers and lakes. Water therefore also cycles through human settlement – supporting life, carrying waste, growing food and returning to its source.

As water is drawn from a well, drawdown increases with distance from the well and then stabilizes at some distance from the well. The area between the normal water level and the pumping level forms a cone of depression (Figure 7.2). The larger the withdrawal, the larger the cone. The ground water will flow towards the well, into the cone of depression. This will alter the direction of flow of groundwater. In some areas, overdrawing can actually result in subsidence of the surface.

Indirectly, human activity also removes vegetation and replaces it with impermeable surfaces which increase and speed up surface run-off.

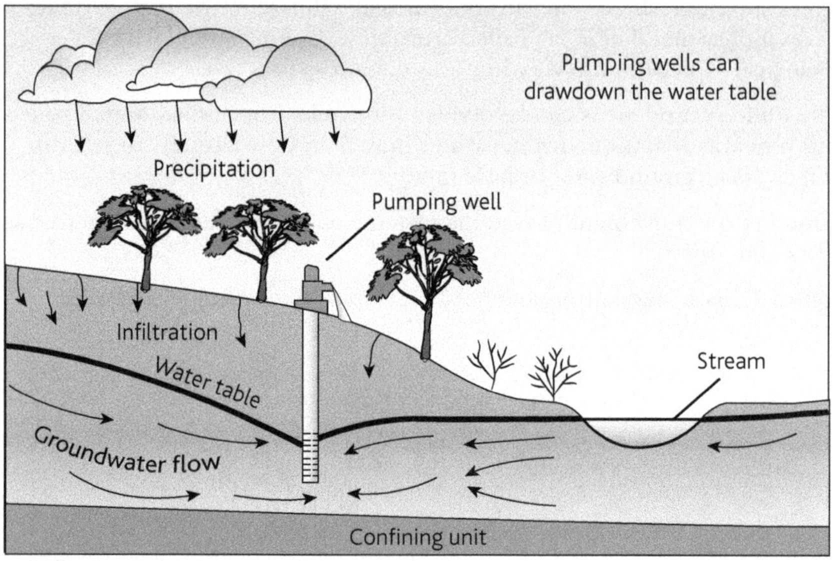

Figure 7.2 *Cross section through hillside, showing through flow, percolation, groundwater and wells with cone of depression, water table and springs*

Activity 7.1

1 Define the terms: interception; infiltration capacity and base flow.

2 Explain the relationship between (i) infiltration and interception and (ii) infiltration and rivers.

3 Explain how human activity can influence the hydrological cycle.

Feedback

1 Interception is precipitation that is trapped by vegetation and prevented from falling directly on the ground.

Infiltration capacity is the maximum rate at which water can enter the soil.

Base flow is the movement of ground water into the river by seepage and interflow.

2 i Infiltration is the rate at which precipitation is absorbed by the soil surface. On vegetated slopes, interception breaks the direct path of precipitation to the surface and slows the rate of its arrival. Up to a third of precipitation may reach the ground indirectly off the vegetation. This slower arrival rate increases the amount of infiltration as previous soil moisture would have had time to move into the soil. Where there is little vegetation, the soil pore spaces quickly become filled by rapid infiltration and the excess runs off the surface.

ii Infiltration is directly related to rivers or channelled surface flow, since rivers represent water that is in excess of infiltration. Where infiltration rates are very rapid, for example, on dry desert surfaces, rivers would be very short-lived as there is little excess left to flow on the surface. Rivers flow where infiltration is at its maximum, for example, humid climates, and there is sufficient excess to create channelled flow/rivers.

3 Human activity can influence many aspects of the hydrological cycle, particularly rivers and groundwater. Water is essential not only for direct human consumption and domestic purposes but for irrigation of food crops and agricultural raw materials; generation of hydroelectric power, as an industrial raw material and coolant; as well as an important mode of inland transport and leisure activities.

Many rivers have been canalized in flood management programmes and soil surfaces replaced by buildings and drainage canals. These measures increase runoff and affect groundwater storage.

Groundwater sources are very severely affected by human pumping of wells which deplete the water table actually depressing its level around the bottom of the well. This is the cone of depression. In some areas such as California, the ground subsides as result of overdrawing of groundwater. Groundwater stores can also be completely exhausted.

Large urban areas with their many impermeable surfaces such as roads, sidewalks and buildings, also impede infiltration of precipitation into the soil. This increases the amount and rate of surface runoff leading to flooding.

Many drainage basins are severely modified by human activity. Rivers are dammed with water collected in large artificial lakes and reservoirs, increasing surface storage. Rivers may also be diverted to supply settlements, irrigate fields and produce hydroelectric power depleting these stores. There are many examples of this effect on the Nile River, Colorado River and the massive Three Gorges project of China.

These are some of the major ways in which human activity impacts on the parts of the hydrological cycle.

The Storm Hydrograph and Water Budget

The river or drainage basin is a very important component of the hydrological cycle. It is the area of land drained by a river and its tributaries. It is separated from other basins by high ground called the watershed. The drainage basin is an open system. Water enters by way of precipitation (input) and leaves as evaporation or stream flow (output).

The storm hydrograph

The storm hydrograph is a graph that allows analysis of river discharge over time after a rainfall event. It is very useful in the prediction of floods, since it plots rainfall in mm and river discharge in cumecs (y-axis) against time in hours/days (x-axis). Flooding is a major hazard in the Caribbean and will be further considered in Module 11, Natural Events, Hazards and Disasters. Figure 7.3 shows the main features of a storm hydrograph.

The precipitation that falls directly in the river channel (channel precipitation), accounts for only a very small proportion of the river's discharge. Most water reaches a river channel after a rainfall event rather than during it. Therefore the river discharge does not start to rise until surface run-off and through flow reach the channel.

The graph shows three distinct sections: the approach segment before the event; the rising limb and the falling/receding limb. The steepness and height of these limbs indicate the likelihood of flooding and the length of time it would take for the river to return to previous discharge level. The lag time between peak rainfall and peak discharge is particularly

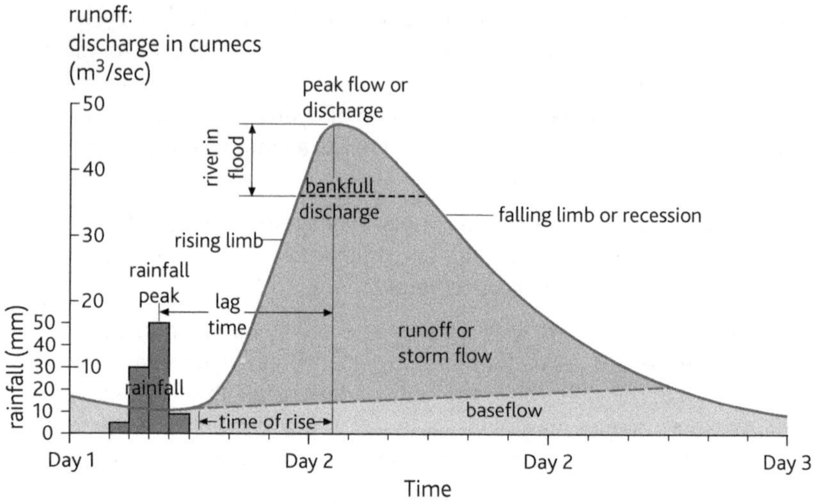

Figure 7.3 *Storm hydrograph*

significant in prediction of floods: a shorter lag time suggests that bank full levels will be rapidly achieved.

The shape of the rising and falling limbs, both in height and steepness, reflects many aspects of the drainage basin. The following are the major influences on drainage basins and hydrographs:

Precipitation

Intense convectional rainfall events such as those which occur in the Caribbean would result in a short lag time and steep rising limb, as the ground is quickly saturated and surface flow enters the channel. Lighter rainfall and short events may result in longer lag times.

Antecedent conditions also affect the lag time: a rainfall event at the end of the rainy season would create a steep, high rising limb as the ground would be saturated, as compared to earlier in the season when more water may be infiltrated.

The type of precipitation is also important. For example, when water is locked up in snow, little water reaches rivers.

Temperature

High temperatures may result in high rates of evapotranspiration and this will reduce discharge. Low temperatures will affect the type of precipitation.

Relief/structure

Drainage basins with steep gradients would have shorter lag times and steeper rising limbs as the water rapidly reaches the channel. Flatter basins would have a more gently sloping rising limb.

In addition, impermeable rocks, such as granite, would promote rapid runoff increasing the height and steepness of the rising limbs. Water passing through pervious limestone and porous sandstones would reach the river more slowly and create a more gentle rising limb.

Soils

Sandy soils allow rapid infiltration. The small pores of clay soils reduce infiltration.

Size and shape of the basin

A small basin would have a short lag time but a lower peak discharge than a larger basin in which water would take a longer time to reach the channel but produce a higher peak discharge. Round basins would have a steeper rising limb with a higher peak, than long basins.

Drainage density

The arrangement and number of the tributaries in the drainage basin would also influence the storm hydrograph. A large number of tributaries in the upper course would result in a short lag time, while fewer tributaries widely spaced would produce a longer lag time. Irregularly spaced tributaries in some basins may even result in two peak discharges as water reaches the channel at intervals.

Vegetation

Vegetated drainage basins have longer lag times and lower peak discharges as the plants intercept and cycle the water, reducing surface run-off and increasing infiltration. The water therefore takes a longer time to reach the channel. Bare slopes have a high short lag time and higher peaks as there is more surface run-off.

Human activity

Intense human activity in a drainage basin can have a profound effect on the storm hydrograph. The abundance of impermeable surfaces, canalized drains and waste water outlets can greatly increase the peak discharge and shorten the lag time as water moves rapidly over these surfaces. Deforestation also affects flows in drainage basins.

Feedback

a i lag time is 8½ hours

ii bankfull discharge is at 36 cumecs and it occurs at 23.30 on day 1.

b The rising limb is steeper than the falling limb indicating that water is rapidly reaching the channel. This could be a small round steep basin so that water rapidly reaches the gauging station. Infiltration and through flow are rapid allowing water to reach the channel relatively quickly after the end of the storm.

The receding limb is often more gentle reflecting the slower process of through flow. The discharge does not return to pre-storm levels until all the storm water has passed through the channel. This may be days later.

Activity 7.2

Using the storm hydrograph given in Figure 7.3:

a Calculate the (i) lag time; (ii) bankfull discharge and time of occurrence.

b Account for the shape of the rising and falling limbs.

Water budget

The soil moisture budget or balance places emphasis on water as a fixed quantity: if it is used it has to be replaced, with a balance between amount received (precipitation) and amount used. But it also takes into account soil moisture storage. A visual display takes the form of a compound line graph, in which monthly precipitation and potential evapo-transpiration in mm are plotted against the months of a year. In a drainage basin, the water balance can be represented by the following equation:

Drainage basin discharge =
precipitation – evapotranspiration +/– changes in storage

The budget varies from one geographical region to another and seasonally. Wetter climates of the Equatorial regions, such as in Guyana, have an absolute water surplus over the year: that is, precipitation exceeds evapotranspiration. Some small tropical marine islands may experience periods of absolute deficit when potential evapotranspiration exceeds precipitation.

Many other areas with seasonal variations in rainfall may be able to use existing soil moisture (soil moisture utilization) to make up the deficit and then replace it with the first rains (soil moisture recharge), until field capacity – the amount held in the soil after the excess has drained away – is reached. Such places may not experience an absolute water deficit at all, or at least not for the entire period of water shortage.

Water budgets could highlight the challenges of river basin management, illustrating when flooding or drought are likely to occur and when irrigation would be necessary.

The varied annual water balance under different climatic conditions leads to varied river regimes. The variation in the flow of a river usually over a year is known as the river regime. Some rivers have very simple regimes and demonstrate a clear difference between one peak and one low. Such regimes may be found in areas with monsoon climates. Others are complex with multiple peaks and lows reflecting differences in relief in parts of the drainage basin or the existence of large tributaries. The discharge of the river changes over the year in response to water availability caused by climate or human interaction. Some rivers may dry up completely during severe water deficit. Others respond with fluctuating water levels. During the dry months rivers may become choked with deposited material which they can no longer transport.

Drainage Patterns and Drainage Density

There are two basic methods of spatially analyzing drainage basins characteristics: descriptive analysis of patterns shown on maps (drainage patterns) and quantitative analysis of streams in a drainage basin (drainage density, stream ordering).

Drainage patterns

Descriptive analysis of drainage basins is an older methodology of classifying rivers. Maps are used to delineate the drainage basin and subjectively determine the visual pattern which it makes. Geological maps are used to explain the formation of these patterns. These patterns which develop where there is significant runoff are influenced by geology and local topography.

Drainage patterns may be classified as accordant, that is influenced by the relief and geology of the areas over which they flow, and discordant or unrelated to local relief and geology.

Accordant patterns

Dendritic

Dendritic drainage patterns (Figure 7.4) are common where the river is flowing over a homogenous surface with little structural variation. Under such conditions there is no control over the direction taken by the streams. The tributaries join the main stream in irregular patterns and at an acute angle, resembling the branches of a tree joining the trunk. Many rivers in the Caribbean show this pattern.

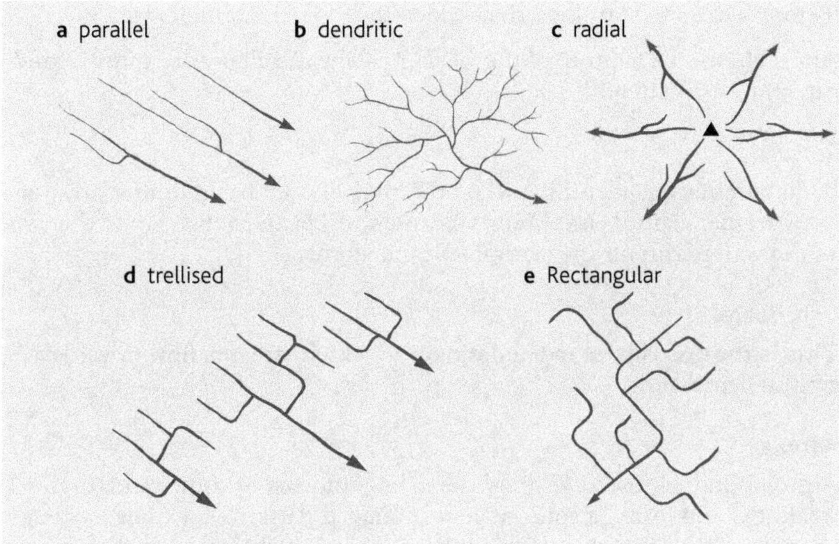

Figure 7.4 *Drainage patterns*

Trellis

A trellis drainage pattern occurs when the streams join each other at right angles. This is the result of structural controls on the rates of river erosion. Softer rock interbedded with more resistant rocks display differential erosion, where the stream on softer rock erodes more quickly and may in fact capture the head waters of the more slowly eroding stream on harder rock. This stream is said to be beheaded. It is then possible to identify the point of capture, the wind gap left where it occurred, the beheaded stream and the subsequent and consequent streams (Figure 7.5). Trellis drainage may develop in folded topography where the main stream follows synclines and tributaries run down anticlines to join the main channel.

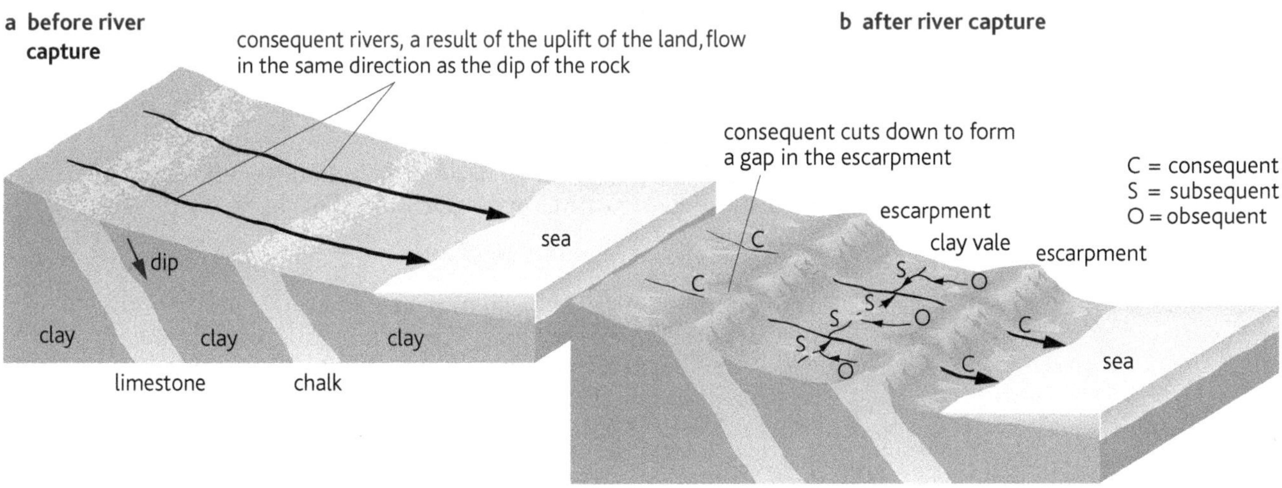

Figure 7.5 *Before and after river capture*

'Before' shows the stronger river widening and extending its valley.

'After' shows the capture of the weaker valley, the elbow of capture, wind gap and misfit stream.

Radial

Radial drainage is identified when the river of each basin drain outwards away from a central rise. Many volcanic Caribbean islands are of this type as the waters run off the central volcanic dome.

Centripetal

This is the opposite of radial drainage in which streams flow towards a central depression.

Annular

Annular drainage occurs where the river conforms to some structural weakness and forms a concentric ring-shaped pattern. The flow is along the rings with tributaries joining them. This is often associated with volcanic domes or other circular structures.

Parallel

This is a simple response of usually newly formed rivers to a single slope. They flow down straight courses following the common gradient. Often very steep slopes.

Rectangular

This pattern looks similar to trellis drainage with streams meeting each other at right angles. However the structural control is different. The river responds to weaknesses such as tectonic faults and rock joints to form this rectangle pattern.

Deranged

Deranged patterns develop as a result of the disruption of a pre-existing pattern. For example, a dendritic pattern may be over run by glaciers which leave material behind which dam streams. The glaciers may contort the tributaries.

Discordant patterns

Some well developed drainage basins show characteristics that are unrelated to the structures over which they flow at present.

Superimposed

River patterns sometimes seem unrelated or even contrary to the structures over which they flow. The patterns suggest that the pattern reflects an older structure. The pattern conformed to original layers which have been eroded, leaving the river pattern discordant to the present surface.

Antecedent

In this case, the river pattern overrides the structural surface. Even as the land rises the river is able to maintain its pattern and vertical erosion. It is discordant to the surface over which it currently flows. The Brahmaputra River in India was able to cut down deep gorges and maintains its original course as the land was being pushed up and folded to form the Himalayas.

Stream ordering

Strahler developed a method of stream ordering based on the number of tributaries (Figure 7.6). A stream with no tributaries, a headwater tributary, is considered a first order stream. Where two streams of order 1 meet the continuing section is order 2. Similarly two streams of order 2 will form a section of order 3, and so continue until all the streams are taken into account. (Streams joining higher order sections do not change the order, for example, order 1 joining order 2 does not make an order 3 stream.) A basin is classified according to its highest stream order.

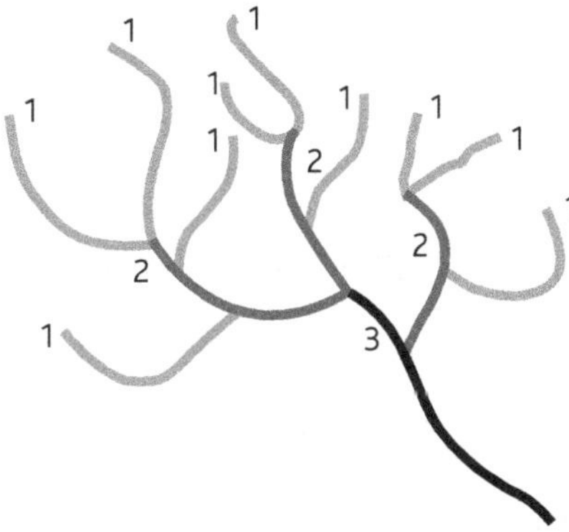

Order	Number of Segments	Bifurcation Ratio
1	10	3.33
2	3	3.00
3	1	

Figure 7.6 *Strahler's stream ordering*

Stream orders of one to three constitute about 80 per cent of the world's waterways. A stream basin of a high order suggests a larger, more complex drainage system with greater length, than a low order one.

Stream orders of seven and more are major rivers. Studies show that there is a:

- Negative correlation between stream order and the number of streams in a drainage basin.
- Positive correlation between stream length and stream order.
- Positive correlation between stream order and the area of a drainage basin. The Mississippi is a 10th order stream and covers about 3.2 million square kilometers. The Amazon River covers about 7.1 million square kilometers and is a 12th order stream.

It was this early work (1952) that allowed a quantitative analysis and comparison of drainage basins:

The bifurcation ratio or the relationship between the number of streams of one order to the next highest order. The higher the ratio, the higher the risk of flooding.

Drainage density

This is found by dividing the total length of all streams by the area of the basin. It is influenced by climate, vegetation, relief, soils and geology.

> **Activity 7.3**
>
> 1 Draw a simplified diagram to represent the 'After' situation shown in Figure 7.5.
>
> Insert: elbow of capture, wind gap, misfit stream.
>
> Indicate a point at which a waterfall may develop.

The Use of Topographical Maps to Identify Factors Influencing Drainage Basin

Topographical maps which show both the relief, vegetation and human activity of an area are very useful in inferring the influence of factors such as geology, vegetation and human activity on the drainage basin.

- Geology can be inferred on a topographical map by examining the break of slope and changes in the drainage pattern itself. The presence of springs and resurgent streams may suggest a very permeable rock type.
- The role of vegetation can be determined by comparing the type and density of vegetation with the drainage of the area. It would be expected that increased infiltration of vegetated areas may result in smaller surface streams than in unvegetated areas.
- The influence of human activity can be easily identified by the straightness of drainage/irrigation canals, by streams disappearing into cultivated areas and the presence of reservoirs or dams.

Key Points

- Water moves between sea, air and land in liquid, gaseous and solid states in a closed hydrological cycle.
- Important flows on land are surface run-off, infiltration, through flow, percolation and base flow.
- The water cycle varies with time and place.
- The storm hydrograph and water budget concepts examine changes with time.
- Factors such as climate, geological structure, vegetation and human activity strongly influence the hydrological cycle.
- Surface run-off in a drainage basin can be described by its pattern and quantified by stream ordering.
- Topographical maps can be used to examine the factors operating in a drainage basin.

Conclusion

The hydrological cycle consists of many flows and transfers. Maps and graphs are used to analyze changes over time and place. Many factors account for the features of drainage basins.

End Test

1. What is the water balance?
2. Explain how climate and vegetation can cause the water balance to vary with time and place.
3. State the formula for the calculation of drainage density.
4. Explain how rock type influences the hydrological cycle.

End Test Feedback

1. The water balance is the equilibrium in a drainage basin between inputs and outputs. It compares input precipitation (P) with outputs of evapotranspiration (E), runoff (Q) plus/minus storage. It can be expressed by the formula P = Q + E +/– storage.
2. Climate is an important factor determining variation in water balance both in time and place. Total precipitation, its seasonality and its form, control the main input. Low total will often result in water deficit, especially if it is coupled with high temperatures and therefore high potential evaporation. High total amounts would create a surplus and significant water storage. Seasonality of precipitation would determine whether rivers are perennial or seasonal. The form of moisture-frozen/liquid – affects its availability for infiltration or runoff.

Vegetation type and seasonal variation have a significant role to play in the water balance. Evaporation and transpiration occur through the leaves of the vegetation. Therefore significant vegetation cover will result in higher evapotranspiration rates. Bare ground or winter leaflessness would result in lower evapotranspiration rates especially if combined with low temperatures. In addition, vegetation influences infiltration as it slows down arrival of precipitation on the surface by interception. Highly vegetated slopes will have more infiltration and less runoff. The vegetation also stores moisture.

3 Formula for drainage density is Density = Total stream length (km)/Area (km^2)

4 Rock type is important in controlling the infiltration and ground water movement and storage.

Impermeable rocks increase runoff and reduce the amount infiltrating and reaching ground water sources.

8 Fluvial Processes and Landforms

General objective

On completion of this module you should develop an understanding of fluvial geomorphic processes.

Specific objectives

You should be able to:

- explain the main concepts, flows and processes associated with rivers
- explain fluvial processes which influence the development of related landforms
- analyze the influence of physical, biotic (human and vegetation) and geological factors on fluvial processes
- apply concepts and processes related to rivers at different geographical scales
- analyze the effect of sea level changes on rivers
- apply map reading skills, appropriate investigative and practical techniques.

Introduction

Rivers are the most widespread and powerful geomorphic agent on land. They occur in diverse environments, paradoxically even in deserts. In Module 7 – Hydrological Processes – drainage basins were examined. This module will look more closely at the fluvial processes and landforms in the valley and channel within the drainage basin.

The channelled water/stream flows from source to mouth, in varied drainage patterns, eroding, transporting and depositing. Many rivers have multiple sources: areas of heavy rainfall, snowmelt, springs and lakes. But the majority flow into the sea under the force of gravity.

This module will examine the channel morphology, the processes of stream flow, and landforms associated with these processes. These will be investigated in plan and long section along the river's course, as well as in cross section.

Content

- Channel morphology.
- Stream flows and processes.
- Valley and landforms.
- Factors influencing fluvial processes, including sea level changes.
- Influence of weathering on river basins.

Channel Morphology

The river channel is formed when surface run-off is concentrated into a long depression. Tiny rills and larger masses of sheet flow coalesce to form the channel. The width and depth of the channel vary along its course. As the water flows downhill under gravity seeking sea level or its base level, the channel reflects the fluvial processes. Near its source, high above base level, dominant vertical erosion produces a channel deeper than it is wide, while nearer its mouth at base level, lateral erosion produces a channel which is often wider than it is deep. In some places, the channel may be asymmetrical being deeper nearer one bank than the other. Deposition also shapes the channel as material is deposited along its course – from large boulders near its source to finest silt near its mouth.

The channel is defined by its wetted perimeter (Figure 8.1). This is the length of the surface of the bed and banks which is in contact with the water in the channel. But channels of different shapes can have the same wetted perimeter. Therefore channel shape is better reflected by the hydraulic radius, which is the ratio of cross-sectional area of the channel to its wetted perimeter. The actual shape of any channel is influenced by rock structure, biotic forces and mass wasting.

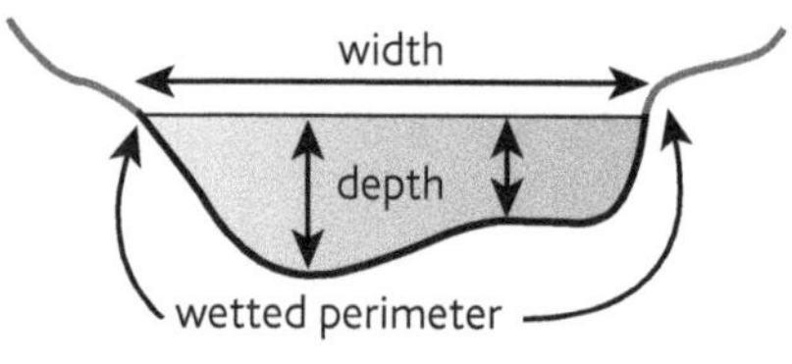

Figute 8.1 *Wetted perimeter*

The nature of the bed and banks influences the type of flow and speed of flow in the channel. Rough surfaces with large boulders increase friction and slow the water, whereas smooth surfaces and fine deposits result in increased velocity. This helps to explain increasing channel velocities downstream of the river. Friction also plays a part within the channel cross section. Highest velocities are recorded just below the surface of the water away from the frictional drag of the bed, banks and air.

Feedback

1 Wetted perimeter – 14m.

2 Example channel 2m wide and 6m deep – wetted perimeter 14m.

3 a hydraulic radius – Area/perimeter – $48m^2/20m = 2.4m$.

  b hydraulic radius –$10m^2/9m = 1.1m$.

4 The larger hydraulic radius in a) suggests that this channel has a lesser amount of its cross sectional water in contact with the perimeter. It would have a greater velocity because of less friction. Stream in b) with a smaller hydraulic radius would be slower as more water would be in contact with its perimeter and be slowed by friction. The wetted perimeters of 20m and 9m, respectively, would only indicate the relative size of the channel but not their shapes. Many different shapes could result in the same wetted perimeters.

Activity 8.1

1 If the channel is 10m wide and 2m deep what is its wetted perimeter?

2 Give an example of other channel dimensions which could have the same wetted perimeter.

3 What is the hydraulic radius of a channel

  a 12m wide and 4m deep?

  b 5m wide and 2m deep?

4 Compare the two channels in 3 above and explain why the hydraulic radius offers a better comparison of the shape of the channel than the wetted perimeter.

Field work

In doing a field investigation of a river, channel dimensions and stream velocity calculations are done at many sites along the course of the river. At least, one measurement is taken at a site near its source and one near its mouth, although it is usual to take these measurements at many sites.

The channel geometry is measured as follows:

- width – measuring tape/rope extended across the channel from bank to bank;
- depth – at equal intervals (1m) across the width, a graduated pole is placed in the channel and depth recorded;
- channel gradient downstream is measured by simple calculations – using ranging poles or clinometers to measure angles of depression at measured intervals downstream;
- rack marks – evidence of variation in water levels along the banks, for example, flattened vegetation – height above current water levels recorded;
- roughness of bed sketched.

The velocity of the stream is measured by timing the movement of a float over a known distance of the course. Sketches and observations are recorded about the type of flow and changes in direction of the river course.

Understanding channel morphology is essential to investigating the work of fluvial processes in landform development.

Stream Flow and Processes

The stream of water in the channel rises and falls according to available water: during dry periods it may be a trickle along the bed, that is, below bank full; in wetter periods it may fill the channel – bank full; under extreme conditions it may even be over bank full; that is, in flood or spate. The amount of water in the river is indicated by its depth and volume.

The seasonal variation in the amount of water in the river is called its regime. Many rivers show variation in the amount of water passing through it seasonally. Most have periods when the source/s contribute more water, for example, spring melt water or storm rainfall – so that the channel may be full or even flood. In other periods the water level falls as the river is maintained only by base flow from groundwater sources.

The velocity of the river is influenced by the gradient of the channel and the volume of the water. Along a tributary or first order stream the low volume, even when flowing over a steep gradient, results in relatively slow movement; while nearer the mouth, the high volume of water in the third or fourth order stream, even flowing over a low gradient, results in faster flowing water. Generally velocity increases downstream as more water is added to the main stream and friction is reduced by the smooth bed.

It is often useful to measure both the volume and velocity of the river together at a given point along its course, this is the river discharge. The place where the measurement is taken is called a gauging station and the measurement is given in cubic meters per second (cumecs).

The water in the channel flows in different ways: turbulent eddying may produce circular water motion, while some water may flow in a straight path producing laminar flow. Laminar flow is rarely encountered in nature and would be found at extremely low velocities, travelling over the stream sediment without disturbing it. There may also be secondary motion in a corkscrew manner called helicoidal flow. Under exceptional conditions it may even free fall over a vertical edge as a waterfall.

The discharge and types of flow of a river at a particular point along its course influence the processes which will occur. Processes of erosion, transportation and deposition occur along the entire course, changing the shape of the channel and ultimately shaping the whole river valley.

Erosion

The main processes of fluvial erosion are:

- *Hydraulic action* – the sheer force of the water impacts the bed and banks of the channel wearing it away. It includes cavitation where air in the rock is compressed and released leading to erosion.
- *Corrasion* – the materials carried by the river have an abrading effect on the channel. In potholing small pebbles caught in the bed swirl around creating deep holes, thereby lowering the bed.
- *Corrosion* – the chemical composition of river water reacts with many minerals, changing their chemical composition or even removing them in solution.
- *Attrition* – the collision of the material in the river with itself accounts for the decrease in size and smoothness of river materials downstream. Large boulders of the upper course are ground to find silt or alluvium near its mouth.

The combined processes of erosion result in the down cutting of the bed of the channel by vertical erosion as well as widening the channel by lateral erosion.

Transportation

The river picks up material which is carried away by it. The material or sediment transported by the river is called its load. Generally rivers with a high discharge have heavy loads, but factors such as the nature of rock type, type of flow and size of particles will influence the nature of the load and the way in which it is carried.

- *Bedload* – the largest boulders and pebbles are rolled (traction) or bounced (saltation) along the channel bed.
- *Suspended load* – the fine particles of clay and silt are carried along in suspension in the water. This accounts for a large part of the total load in the river basin and gives the water its colour.
- *Solution load* – minerals are transported in solution in the river.

The competence of the river water is defined as the maximum size of material the river can transport. Rivers that are in flood can move boulders and pebbles. They also transport large quantities of material. The capacity is the actual load of the river. This would depend on the specific conditions in the channel, for example, size of particle available to be transported. The relationship between particle size and river velocity is shown on the Hjulstrom curve, Figure 8.2. It is useful in understanding the dynamic interaction between the processes of erosion, transportation and deposition in the river channel.

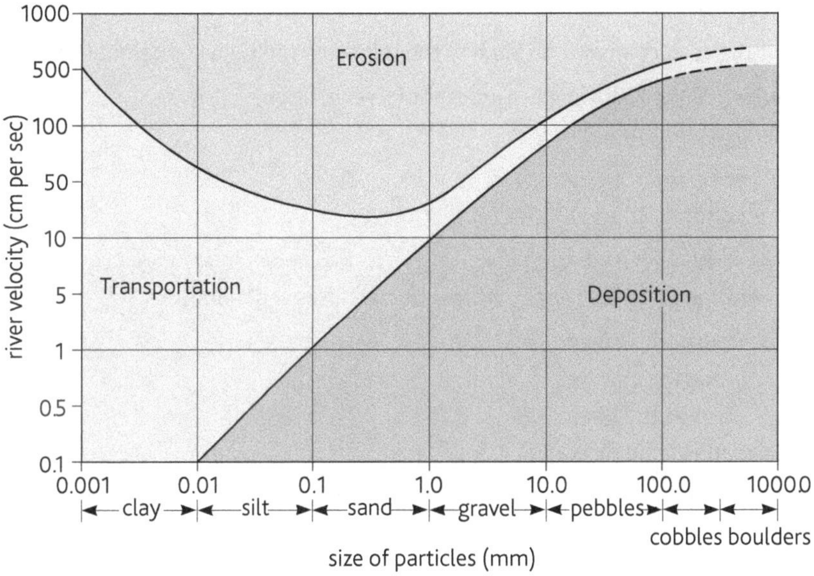

Figure 8.2 *Hjulstrom curve*

Activity 8.2

Use Figure 8.2 to answer the following questions:

1. What is the mean or critical erosion velocity curve?
2. What is the lowest velocity at which sand is deposited? (Investigate the six limits of soil fractions.)
3. Describe two significant features of the relationship between velocity and particle size shown by the graph.

Feedback

1. The mean erosion velocity curve shows the speed at which particles are picked up or eroded by the river. For example, sand particles will be eroded at lower speeds than either clay or pebbles.
2. The lowest velocity at which sand is deposited is 1 cm/sec.
3. One feature is the fact that clay can be transported at very low speeds of 0.1cm/sec, although they can only be picked up at high speeds up to 500cm/sec. Another feature is the very narrow range of velocities at which the largest particles, such as boulders can be carried. A small drop in velocity would result in the largest particles being deposited.

Deposition

The relationship between transportation and deposition is a very dynamic one. The load of the river is deposited when the volume or velocity of the river is decreased. The volume of water may be reduced by a dry period, change in depth or passage over a permeable surface; while the velocity can drop at a change of slope, entrance to a sea/lake or flooding outside the channel.

The load is moved along in successive cycles of being picked up (entrained) and put down (deposited) with fluctuating water conditions. Within the cross section of the channel, the inner bend would have deposited materials in response to slackened speeds.

As the river's competence and capacity change the large boulders are deposited first, while finest clay and silt may remain in suspension along the entire river course. These clays may not settle out until flocculation occurs, when fresh water and salt water meet at the mouth of the river. As the river approaches its mouth, it deposits its load often forming a flood plain and a delta. This creates new land in the sea.

Activity 8.3

Select the correct answer on each of the following multiple choice questions.

1 Which of the following is a process of river erosion?

a saltation
b river cliffs
c corrosion
d infiltration.

2 Corrasion is the process by which the river uses sediment to:

a transport alluvium
b erode the channel
c dissolve the load
d deposit chemicals.

3 Sliding of particles along the river bed is called:

a traction
b suspension
c saltation
d gliding.

4 What type of flow is seen as smooth flow parallel to the river bed, with little or no mixing?

a current
b turbulent
c orthogonal
d laminar.

5 The competence of a river is the:

a efficiency in eroding materials of different sizes
b maximum size of particles which it can transport
c total load it can carry at a given discharge
d effect of the load on the graded profile.

Feedback

1 c   2 b   3 a
4 d   5 b

Valley and Landforms

The river valley is the long depression in which the river channel is found. However, rivers may also occupy valleys created by other processes, for example, faulted valleys. In addition, some river valleys are influenced by geological conditions to form steep-sided gorges along sections of their courses.

In newly formed rivers, the water flows over the irregular surface of, for example, the emerging volcano or uplifted fold mountain. It will flow under gravity seeking the most direct path to its base level. Over time vertical and lateral erosion will wear the valley down and back, changing the initial V-shape into a more flattened U-shape. At first the channel will occupy the entire narrow valley floor while later on the channel may wander over a wide alluvium-covered valley floor.

Activity 8.4

1 Distinguish clearly between the 'drainage basin'; 'valley' and 'channel' of a river system.

2 Describe and explain two ways in which the relationship between the valley and the channel changes over the course of the river.

Feedback

1 The drainage basin is the largest of the three, as it includes all the land drained by the river and its tributaries. This land surrounds the valley up to the divide which separates it from other drainage basins. The valley is the long depression in which the river flows. It includes the land beside the channel and the channel itself. The channel is the actual water-filled depression of the river, consisting of bed and banks.

2 Initially the river channel is very small and occupies the lowest part of the valley. It flows along and is contained in the narrow valley floor, bending around any obstacles. Later on nearer the mouth, the channel is much larger and flows on only a part of the entire wide valley floor.

The channel at first is flowing directly on the rock surface of the valley but later it flows on its own deposits above the valley floor. The rock surface of the valley floor is masked by alluvial deposits of repeated flooding by the river.

The processes of erosion, transportation and deposition are responsible for the creation of many varied landforms within the channel and valley. Although the valley is commonly divided into upper (near source), middle and lower (near mouth) sections, it is more useful to consider these landforms as relating to time rather than place. A recently formed river may show 'upper' course characteristics along its entire course, for example, rivers which flow into the sea as waterfalls. Very old rivers may have 'lower' course characteristics along almost their whole length.

Erosional landforms

- Interlocking spurs are ridges of higher land that jut across the valley on alternating sides, they may be separated by small tributary valleys. The river initially bends around these obstacles. The valley is usually V-shaped and the channel very small.
- Waterfalls are powerful sheets of free falling water which erode the base of their overhang forming a deep circular plunge pool, before continuing on their course. Waterfalls occur when:
  1. a resistant rock, for example, a volcanic dyke, lies across the channel and is eroded more slowly than the softer rocks downstream;
  2. a drop in base level will result in a waterfall from the point (knick point) at which the river starts to adjust to the new base level.

Eventually the retreat of the waterfall upstream will lessen the gradient to form rapids and later graded flow along the long profile. The retreat of the waterfall and its plunge pool upstream as the cliff is undercut and collapses form a gorge.

In the Caribbean, many rivers have waterfalls along their courses as a result of differential rock hardness or uplift. The Kaieteur Falls on the Potaro River in Guyana is the highest but many others exist in the geologically young Caribbean islands such as the Dunn's River Falls of Jamaica. The Kaieteur Falls, five times higher than the Niagara, plunge 251m over the plateau edge forming a gorge as it retreats upstream.

Features of erosion and deposition

- Pool and riffle sequences (Figure 8.3) of erosion and deposition along the channel occur when the river has increased in volume and is transporting sediments as it starts to bend. In the crossover area of the channel between bends, shifting bars result in shallow water which is the riffle. The outer bank is subject to erosion and undercutting forming a deep pool while on the inner bank deposition is occurring.

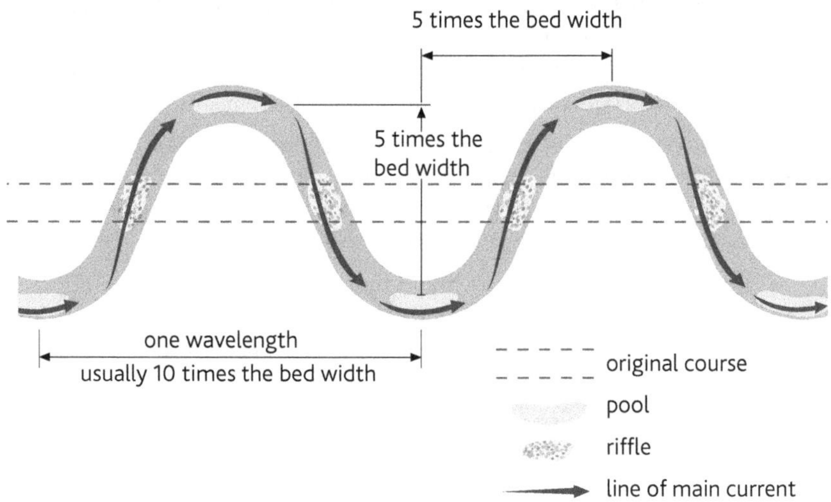

Figure 8.3 *Pools, riffles and meanders*

River cliffs (Figure 8.4) are produced as the bends become more pronounced.

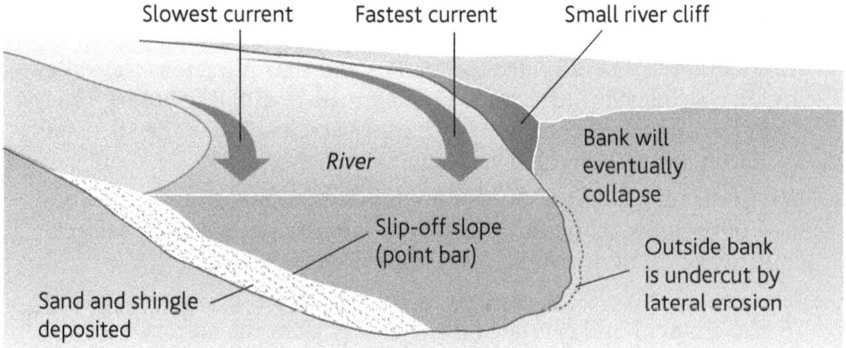

Figure 8.4 *Formation of river cliff and slip off slope*

The undercutting of the bank at the pools results in the formation of river cliffs/bluffs on the outer bends which widen the channel. On the inner bank of the bend the current is slack and deposition occurs on the slip off slopes.

- Meanders and ox bow lakes are best developed after the river has worn back the spurs and is alluvium-laden. The joining of the point bars results in a sediment covered, wide valley floor. The gradient of the channel is very low and the river wanders in great loops over the valley floor.

There are many hypotheses for the initiation of meanders including deflection around an obstacle:

- Variations in velocity downstream associate with the formation of pools and riffles; secondary flows across the dominant downstream flow, but none have been found to be applicable in all cases. However, laboratory simulations confirm the sinuous pattern. It has been suggested that this is perhaps the most efficient form for water and sediment discharge.
- The meanders form and re-form over time, migrating over the valley floor, cutting off meandering loops as ox-bow lakes. The ox-bow lake is a meander which becomes cut off by erosion on the outer banks until the neck of the loop is breached and the meander is bypassed. Deposits seal it off from the main river. These crescent-shaped lakes are temporary and their dried beds are evident on the valley floor. Meanders migrate over the floodplain in this way.

Depositional landforms

- *Point bars* are formed on the inside of bends in the river. The current is slower on the inside of the bend the river therefore cannot carry its load which is deposited. This results in the shallow 'riffle' in the pool and riffle sequence.
- *Braiding* can also be considered a channel form as the river blocks its channel with deposited material around which it has to flow. It occurs when the river is overloaded with sediment so it deposits islands or eyots around which smaller streams flow. Once its capacity is restored the water flows away in a single channel. This can happen in response to changes of velocity or volume or even the nature of material over which the river flows. Rivers which have highly variable discharge and banks which are easily eroded form braided channels. Braided channels in plan look like a bulge in the course of the channel, with many small channels interweaving each other.

  Braided channels are very unstable, migrating laterally in periods of high discharge. This is reduced if the islands become vegetated with their roots adding cohesion to the sediments.
- *Flood plains and levees* are common landforms of large river systems. Rivers with heavy loads and maximum discharge as they approach base level periodically overflow their banks, spreading water and silt over the wide valley floor. These sediments build up until the river channel is often higher than the valley floor and flows on its own deposits. This is the gently sloping, alluvium-covered plain over which the river meanders.

  Levees are the natural banks on either side of the channel formed from the initial deposition of the largest material carried by the flood water, nearest the channel itself. Finer particles are spread further away up to the *bluff line*.

These are common features of the world's largest rivers such as the Mississippi and the Nile.

- A *delta* is the final landform resulting from the river entering the sea and having lost its competence. The load is deposited in different layers or *beds*. Nearest the mouth of the river the coarsest sand form the *topset* beds; then the *foreset* beds of silt and clay extend outwards at an angle; and finally the finest clays are carried in suspension until they coagulate by flocculation and are deposited as *bottomset* beds. The deltas are very flat and are at, just below or just above sea level. There may be one main channel and scores of distributaries branching out throughout the delta and reuniting, that is, *anastomosing*. Deltas are generally triangular in shape but may be:

  Arcuate deltas which are rounded with convex outer margins, for example, the Nile, Rhone and Danube deltas. These form when, in spite of the fact that the process of deposition is stronger than erosion, waves determine the shape of the delta. Where the coastal shelf is steep, waves have the energy to attack fluvial sediments and long shore drift redistributes sediment along the coast.

  Cuspate, or tooth shaped form when a river enters a straight shoreline with strong waves which force sediments to spread out evenly on both sides of the one major channel. This creates a feature with a pointed tooth shape, for example, the River Tiber in Italy.

  Bird's foot: during floods a river may abandon its old course and find new outlets – the process of *avulsion*. The River Mississippi has changed course several times. Each lobe of the Mississippi delta represents a different avulsion event. Distribution channels form the 'toes' of the foot. Such deltas develop in areas where there is a low coastal shelf gradient and wave and tidal action are virtually absent. The river therefore determines the shape of the delta.

  For deltas to form, high discharge rivers must have a high sediment load, and flow into relatively calm offshore areas with a low gradient.

Alluvial fans

These are fan shaped alluvial deposits formed when the velocity of streams is abruptly decreased such as:

- At the foot of a mountain.
- In areas of variable stream discharges such as in deserts subject to flash flooding, or glaciated regions where there is snow melt in summer.

The surfaces are slightly convex with the gradient and deposits decreasing away from the apex. Their shape causes water to spread widely and rapidly and flooding is often a hazard.

Estuaries

These are depositional features at the mouth of rivers. Many became estuaries as a result of changes in sea level. Tidal creeks and channels alternate with mudflats which are covered at high tide. In temperate regions, grasses may colonise the mudflats and these may be replaced in tropical areas by mangroves.

Activity 8.5

Choose a stretch (about 2km) of a river accessible to you, or a river for which you have a large scale map, for example, Port Antonio, Jamaica.

From your knowledge of rivers, state the purpose of your activity. Then do the following:

1. Select at least three observation sites along the course, locate and number the sites on a map.
2. At each site:
   a. measure and record the width, depth, velocity and type of flow of the stream;
   b. sketch the cross sectional shape of the channel, banks and flood rack marks;
   c. locate in plan, landforms such plunge pools, river cliffs, point bars. If possible measure and record their dimensions;
   d. sketch the valley including any deposited material, human activity, vegetation, rock outcrops and mass movements and weathered materials;
   e. measure and record the downstream gradient;
   f. collect water and sediment samples;
   g. note any unusual or significant characteristics which may later assist in analysis of the processes at the site.
3. Using additional textbooks and studies, account for the valley and channel characteristics described and measured.

Feedback

(this is a sample of the required presentation)

Purpose: To analyze the processes and factors shaping the characteristics of the valley and channel of river *x* from the bridge at ford to the crossing at highway 3.

	Width	depth	velocity	flow
Site 1				
Site 2				
Site 3				

Notes and sketches

Analysis

Sample: the narrow channel with its turbulent flow appeared to be vertically eroding its bed. Potholing was seen in the bed. Large boulders present suggest that the river has a greater competence than was observed. The shallow water could not transport material of that size. It is also possible that the boulders originated from the collapse of the banks which had similar boulders embedded in them.

Long profile

Streams need gradient to transport their bed load and to do this, a stream has to adjust its gradient. The gradient of the channel changes along the course of the river from its source to mouth. These changes are reflected in its long profile. From headwaters, the gradient diminishes continually. In the upper reaches, discharge is small. A small stream must have a steep gradient to transport its sediments. As the discharge of the stream increases, the channel cross sectional area and the hydraulic radius increase. Because of this increase, there is less friction. The stream becomes more efficient, particle size decreases and the stream can do its work on a lower gradient. So the gradient decreases. When the supply of the load matches the capacity to transport it the stream is said to have achieved a graded profile.

The ideal graded long profile of a river is concave upwards, steep near the headwaters and flatter near the mouth of the stream. Few rivers attain the graded profile and if they do, it is not maintained for extended periods because of changes in climate and changes in sea level.

Factors Influencing Fluvial Processes

The factors influencing the river drainage basin have been discussed in Module 7 – Hydrological Processes. Climate, geologic, vegetation and biotic factors each impacted on the hydrograph and the drainage patterns. These factors also influence the processes and landforms found in the valley and channel.

Weathering is the break down and decomposition of rocks by atmospheric conditions. It does not involve any transport of material and is therefore not an agent of erosion, but one of the important processes of denudation. The weathered materials will accumulate where they occur or fall/slide/flow/creep under the pull of gravity. The movements of weathered materials, with or without water or ice, are called mass movements.

After a river channel has been initiated, all the valley areas above the water are subject to weathering. Depending on the climate, rock type and structure and vegetation, these weathering processes may reduce the strength of the rock material and, especially under conditions of heavy and prolonged rainfall, provide additional load for the river to transport; control the steepness of the valley sides and alter the overall drainage basin characteristics.

Slope stability also affects fluvial processes. To some extent this depends on the geology and soil. Some slopes consist of unconsolidated materials which may slide off into river channels precipitating flooding. In clays, increasing moisture may cause the slopes to fail. Hard rocks are more stable. They often form slopes of great steepness and height which can fail spectacularly. These processes deliver extra sediment to river channels which causes erosion and accelerate deposition downstream.

People have always been attracted to the fertile floodplains of rivers and it was here that the great civilizations of the world developed – the Egyptians on the Nile, the Sumerians on the Tigris and Euphrates rivers. They were attracted to these areas because of both the presence of water and the agricultural potential of the floodplains. River terraces also provided fertile agricultural land. Today, fluvial landscapes are valuable assets for tourism – the Niagara Falls, the Grand Canyon.

The activities of man have impacted on fluvial processes. Vegetation stabilizes slopes and this is particularly important in steep headwater areas. Forest removal, the grazing of animals result in gullying. In addition, the drainage network may be extended in the upper reaches of the basin. Urbanization has also increased the supply of sediments to stream channels, aggrading and eroding. Urbanization may increase surface runoff leading to a rise in peak flows.

Any change in the supply of sediments to streams will alter the character of the streams. Meandering may be transformed to braiding.

The current period of global warming may result in a rise in sea level as ice caps melt. This would result in drowning of the lower course of the river valley to form bays and headlands or rias and islands according to the alignment of the valley to the coast. Along the long profile the main process would be deposition as the river seeks to flow at the raised base level.

A lowered base level as during glacial periods or rising land will result in more erosion as the river grades its profile. Changes in the dominance of erosion over deposition from the height at which the base level dropped is called the knick point. This is often marked by waterfalls where the river flows to the new base level. As the river cuts down into its own previously deposited sediments it forms terraces either side. Former meanders are cut into V-shaped valleys but maintain the sinuous patterns as incised meanders. The river is said to be rejuvenated along this section.

Key Points

- Channeled water is an important agent of erosion.
- Fluvial processes shape the landscape producing many landforms.
- Physical and human factors influence the effect of these processes.
- Weathering is effective in river basins.

Conclusion

Rivers are powerful agents in shaping the landscape. They change as they flow over the land. Processes of erosion, transport and deposition create many significant landforms such as waterfalls, floodplains and deltas. Changes in sea level keep them in dynamic flux as they seek their base level.

End Test

With the use of labelled diagrams, show how a river can alter the shape of its channel through erosion and deposition.

End Test Feedback

The shape of a river channel in both cross profile and plan is significantly altered by fluvial processes. Erosional processes widen and deepen the channel in order to flow at base level; while deposition smoothens out any irregularities along the course.

The small V-shaped channel which is initially formed, is deepened by potholing and other processes of vertical erosion. Bed load also scrapes the bed as it is transported. So from a few centimetres in depth, the channel near its mouth can be many hundreds of metres deep.

As the river increases in volume and velocity, its bending results in undercutting on the outside of the bend where the flow is strongest. This widens the channel until the valley sides are worn back to low bluffs. The channel may now have a wider U-shape.

At the same time, channel load is not only abrading the channel but also being deposited along its course. The large boulders often found near its source gives the channel a rough and irregular bed. Further downstream as attrition has reduced the particle size, changing velocity and volume may result in the channel being covered in alluvium and in fact flows on its own deposits.

9 Coastal Processes and Landforms

General objectives

On completion of this module you should:

- understand the geomorphic processes which shape coastal environments
- be able to use appropriate skills and techniques.

Specific objectives

You should be able to:

- explain the main concepts, flows and processes associated with the marine environment
- explain coastal processes which influence the development of related landforms
- analyze the factors which affect the processes operating within coastal areas
- explain the effects of human and physical interaction within the coastal area over time
- apply concepts and processes related to coast areas at different geographical scales
- analyze the effect of sea level changes on coastal landforms
- apply map reading, appropriate investigative and practical techniques.

Introduction

The coastal environment is a very dynamic one. It responds to daily tidal changes, fluvial, weathering and wind processes, as well as wave action. In the Caribbean, coastal activity dominates not only the islands but mainland countries along the Caribbean Sea and Atlantic Ocean. The importance of this environment goes beyond its geomorphic significance, as Caribbean coastal environments are heavily populated and are important economic resources for tourism, fishing and transportation. These activities have a strong influence on the physical geography of the coast.

This module looks at the processes by which waves shape the coastal environment by erosion, transportation and deposition, forming distinct landforms such as cliffs and beaches. It also examines coral reef coastal environments.

Content

- Wave formation, structure and type.
- Major flows and processes.
- Landforms produced by erosion, transportation and deposition.
- Formation and distribution of coral reefs.
- Geological and human factors and coastal landforms.
- Influence of sea level changes.
- Topographical maps, aerial photographs and field sketching.

Wave Formation, Structure and Type

Waves are the dominant agent of erosion in coastal environments. Most waves are created by winds disturbing the surface of the water. Wave action is strongly influenced by tidal changes and winds.

Tides

Tides are the coastal fluctuations in the level of the water: in fact, the coastline can be said to migrate seawards and landwards twice daily. At high tide, a greater portion of land is under water than at low tide when more land is exposed. The result is that the area of actual land/sea contact, and wave action, varies twice daily, about 12 hours and 25 minutes apart. These changes are caused by the gravitational effects of the moon and the sun on the earth's water.

The difference in height between the high tide and low tide is called the tidal range. In the Caribbean, the tidal range is relatively small, but can amplify the effects of storm waves. In other places, such as the Bay of Fundy, in eastern Canada, it can be as high as 10m. Tides may also vary with the phases of the moon. The higher spring tides occur during full and new moons and lower neap tides, during the quarter moons. They reflect the changing sun/moon alignment.

Other unusual conditions can push the water out to a level below that of low tide (before tsunamis) or above high tide (storm surges, tsunamis). Generally the average level between high and low tide is called the mean sea level. This is not a constant as the level of the land relative to the sea is dynamic resulting in emerging/submerging coastlines. Sea level also changes over geological time as more or less water is in the ocean basins. The current period of global warming is resulting in rising sea levels.

Fetch

The wave fetch is maximum distance of unbroken water over which the wind can blow. A long fetch produces high energy waves of great height. The fetch of the waves on the Atlantic east coast of Caribbean islands, can be as much as 3000km; while those on the Caribbean Sea have a fetch of less than 1000km. The prevailing North-east trade winds and tropical storms also approach from the east resulting in the generally rough seas on the Caribbean east coasts as compared to the calmer waters of the west coasts. On the Trinidad west coast and Jamaica north coast, the winds travel less than 100km from Venezuela and Cuba respectively, resulting in low energy waves.

Structure and types of waves

Generally, a wave is a disturbance of the surface of the water by wind. The water rises in crests and goes down in troughs. There are many different types of waves, but waves are described by their:

Height (H) – distance from trough to crest (separate from the depth of the water)

Length (L) – distance from crest to crest

Period/frequency – time taken for successive waves to pass a point

Steepness – H/L. this cannot exceed 1:7 because the wave will break at that point.

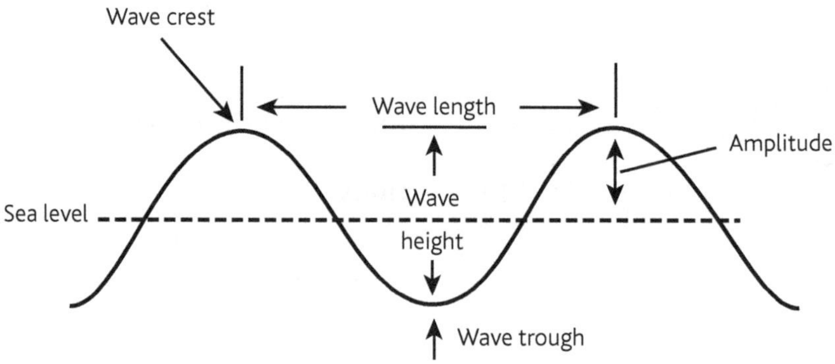

Figure 9.1 *Wave dimensions*

There are many ways of distinguishing between types of waves:

Waves of oscillation and waves of translation

- Waves of oscillation are those in which the water particles move in a circular orbit, but without horizontal motion. These occur in deep water.
- Waves of translation have forward motion as in breakers on the shore.

Swell and sea waves

- Swells are rounded waves of low height, which travel outward from a storm centre or ocean surfaces after wind dies down.
- Sea waves are waves travelling over shorter distances and having greater height.

Breakers-spilling and plunging

- When waves approach shallow water (depths less than half the wave length), friction on the sea bed slows the base of the wave until the wave crest moves past the base and 'breaks' on the shore. On gentle beaches waves may spill. When the gradient of the sea shore is steep or there is a sudden change in depth, and where there is little or an off shore wind, steep wave crests will plunge to shore.
- The forward motion of water on the shore is called swash and is in the same direction as the wave crest; while the return of the water to the sea under gravity is the backwash which is at right angles to the slope.

Constructive and destructive waves

Waves (breakers) may be classified according to the resultant effect on the shore.

- Constructive waves are flat and gentle. They are of heights less than a metre and with lengths up to 100m. They spill over and are low in energy. Stronger swash than backwash results in material being left on the shore (Figure 9.2).
- Destructive waves are taller waves of shorter length. They plunge to shore with high frequency and high energy. The result is a backwash often stronger than the swash removing material from the shore.

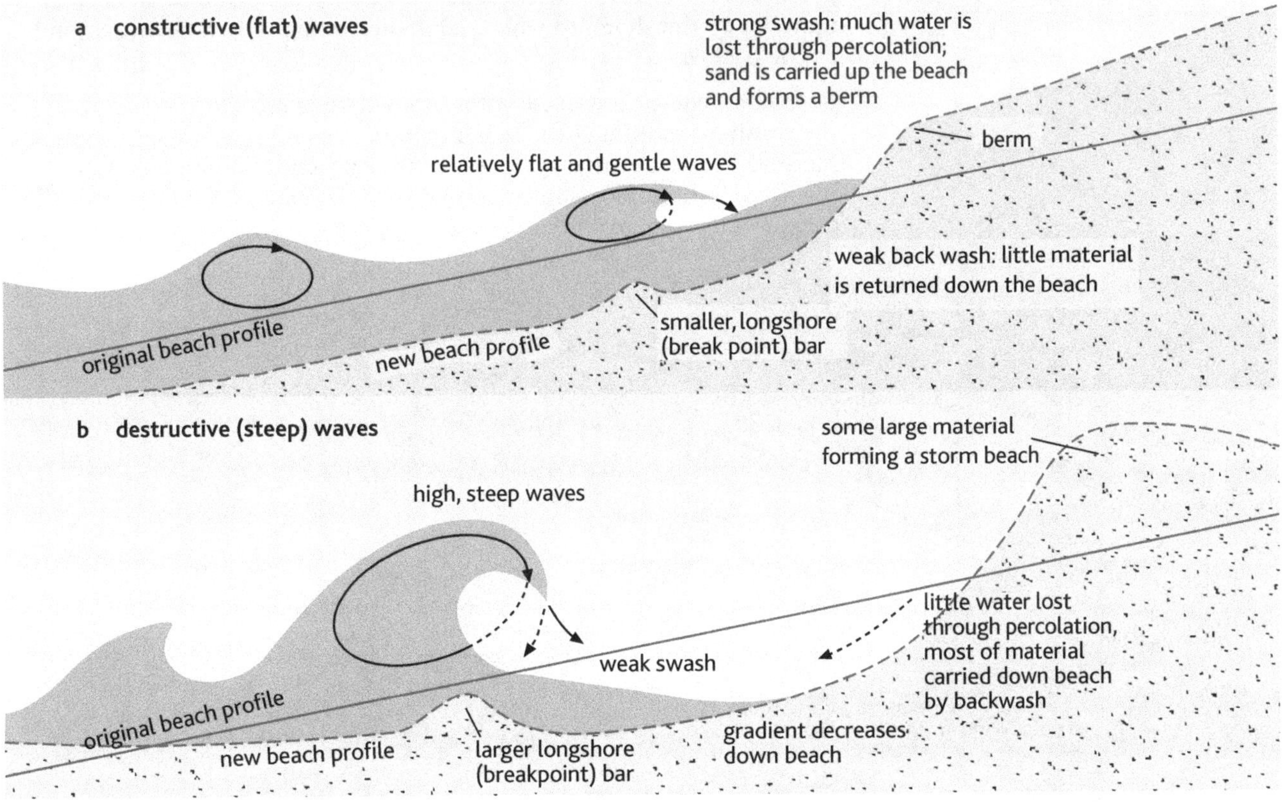

Figure 9.2 *Constructive and destructive waves*

Flows

When viewed in plan there are many different movements or flows taking place on coastal areas.

Currents

On varying scales, there are currents. These are horizontal movements of water on or below the surface. The Caribbean experiences the warm water of the global North Atlantic current moving northwards from the Equator. On a local scale, irregular off shore profiles may result in small currents, for example, rip currents, along the shore.

Wave refraction

As waves approach a coastline their crests may bend to conform to the shape of the coastline. This is called wave refraction. As the wave crest moves on to an embayed coastline, the crest will pile up quicker in front of the headlands than in the bays. The crest will move more slowly over the shallower water in the bay. This differential in speed causes the wave crest to bend parallel to the coast. More importantly, it concentrates the wave energy on the headland, while weaker waves spread out along the bay. Continuous deposition in the bays and erosion of the headlands will eventually smooth out the coastline in plan.

Longshore and beach drift

The process of movement of material both on shore and in the near shore area is referred to as littoral drift.

In beach drift, breakers wash up on the shore as swash, while gravity returns the water to the sea as backwash. The forward movement is parallel to the direction of the wave, which picks up and carries beach material up the shore; the material is then dragged back down seawards by the backwash.

On coasts where the wave front is at an oblique angle to the coastline, the resultant motion of the beach material is along the coast in a zigzag motion. This is the major process of transportation of materials along the coast. A similar process near shore results in material moving along the coast as longshore drift (Figure 9.3).

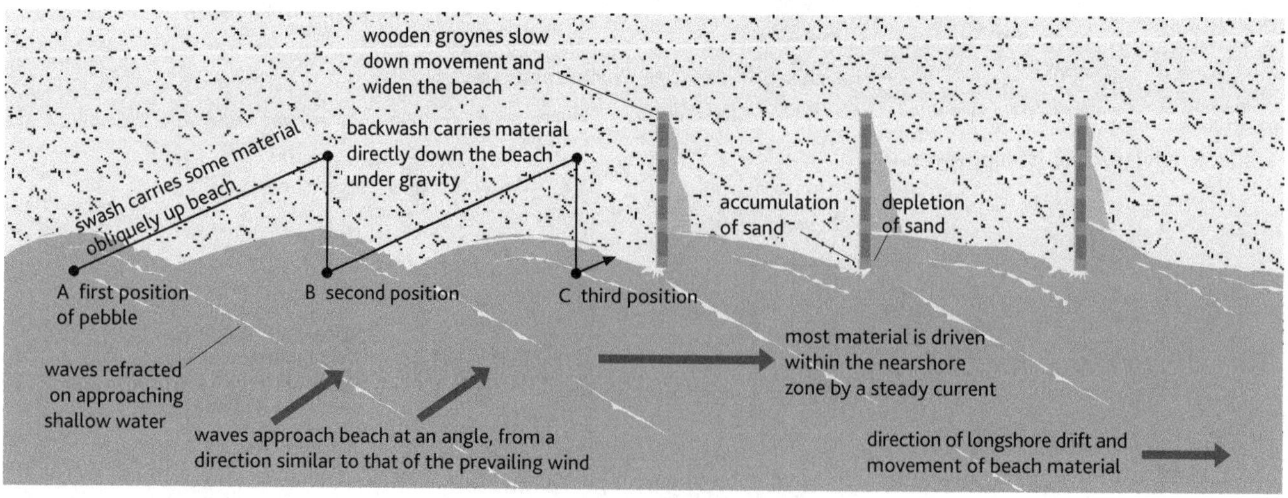

Figure 9.3 *Longshore drift*

Where wave crests are parallel to coastline, the beach material is simply raked landwards and seawards with each breaker. There is no horizontal movement of the beach material.

Feedback

1 Waves are formed by the transfer of energy from winds blowing over the surface of the sea. This disturbs the surface as the motion is translated through the water. Winds also aid in piling up the water on the shore.

2 Sea waves are waves which travel over a shorter distance and have a greater height than swells which are low and rounded waves generated by weather systems.

3 Destructive waves have a higher frequency than constructive waves often breaking at more than 5 per minute. Destructive waves also are steeper and taller plunging to shore as compared to the low spilling action of constructive waves. Finally destructive waves have a stronger backwash which removes material from the shore; while constructive waves have a stronger swash moving material onto the shore.

4 Wave refraction is the bending of the wave crest to conform to the coastline. It is particularly noticeable on embayed or indented coastlines. The water in front of a headland becomes shallow first causing the waves to break at that point with great frequency. This concentrates the wave's energy on the headland. In front of the bays the wave crest continues to move to shore not breaking until it is closer to land as the water gets shallow over a longer distance. The wave energy is dissipated along the bay.

Wave processes are strongly influenced by wave refraction. In the bays deposition is dominant as the waves energy is lessened by the slower movement along the bay; while wave erosion results from wave energy being concentrated on the headlands.

Activity 9.1

1 How are waves created?

2 What is the difference between sea waves and swell?

3 How do destructive waves differ from constructive waves?

4 What is wave refraction? How does it affect wave processes?

Processes

Erosion

Wave pounding

The most dramatic and obvious wave action is that of erosion. The power of large plunging breakers crashing to shore is visible. It also makes surfing, that is the harnessing of that power, such an exciting and dangerous sport. Breakers can exert severe forces on the shore.

Hydraulic action

The breaking wave compresses the air in rock fractures and decompresses as it retreats. This repeated compression and release create stresses in the rock which break them up.

Corrasion/abrasion

The waves carrying material act as a file to wear away the land.

Attrition

The continual motion of every breaker grinds down coastal material to fine sands. The materials jostle against each other, resulting in the characteristic rounded shape of beach pebbles.

Solution

Exposed coastal rocks which are susceptible to chemical action are affected by the sea water.

Deposition

Wave deposition occurs when wave energy is low and waves can no longer carry the materials being transported. This occurs in bays where wave energy is refracted onto the headlands. Deposition is also present when the wave backwash is weak allowing materials carried by the swash to accumulate on the shore. The presence of large rivers along the coast can accelerate wave deposition as they empty their heavy loads of alluvium into the sea.

Landforms

Coastal landforms may be divided into those produced mainly by erosion and those influenced by deposition.

Landforms of erosion

Cliff/wave-cut platform/notch/cave

All the processes of erosion attack the land as the wave breaks between the low tide and high tide levels. Erosion forms a groove in the land called a notch. It is visible at low tide and covered at high tide. It can be traced along the coastline. Previous sea levels can sometimes be traced in notches high above the present sea level.

The waves continue to erode the land until the notch is enlarged to a cave. Joints in the top of the cave may be enlarged to allow water to surface at the top of the land as a blow-hole.

The cave in the side of the land is enlarged upwards by the breakers undercutting the land, until the overhanging rock collapses. This leaves a vertical face or cliff. The gently sloping rock surface of the cave floor is exposed as a wave-cut platform. The rubble from the collapsed overhang may cover the platform.

Throughout the Caribbean, cliffs occur in a variety of locations: raised limestone cliffs of eastern Barbados; resistant cliffs of metamorphic rocks of northern Trinidad and the volcanic cliffs of St Lucia. The angle of the cliff face, its rate of retreat and profile details are all controlled by the geological structure of the area.

Arches/stacks/stumps

These minor landforms (Figure 9.4) occur where wave refraction concentrates wave erosion on both sides of a headland. The back of the caves eventually join, allowing water to pass from bay to bay. This opening may be enlarged to form an arch. Over time, the arch collapses leaving a column of rock separated from the mainland, this is a stack. Undercutting and weathering may reduce stacks to a stump. The location of stacks facing the headland, and their connection to the mainland by a wave-cut platform, distinguish them from other offshore islands. This process aids in the retreat of the headland.

(Islands may be formed by other processes, for example the small islands off western Trinidad are formed by inundation of faulted valleys.)

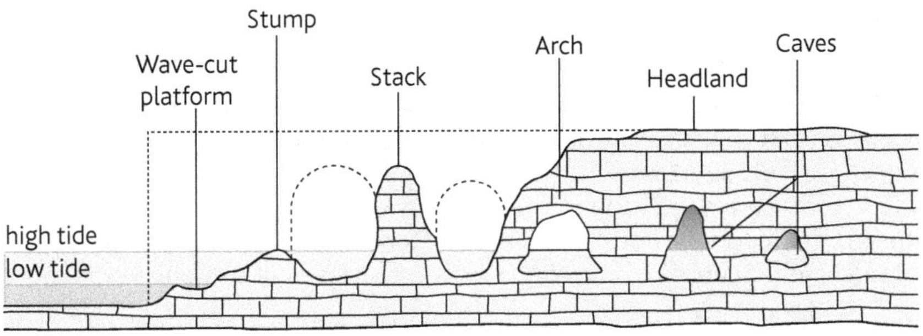

Figure 9.4 *Coastal erosional features*

Depositional landforms

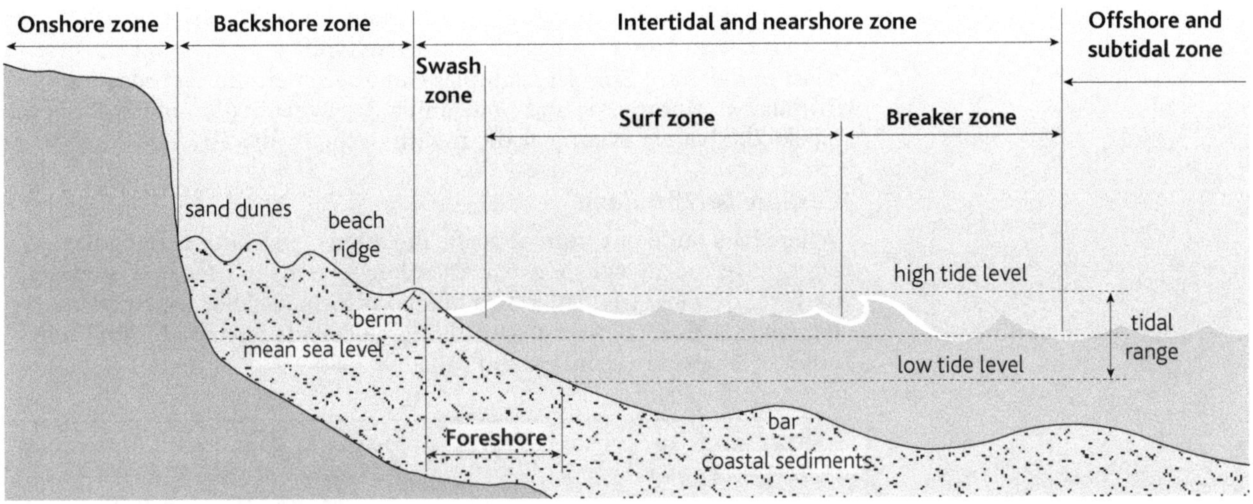

Figure 9.5 *Coastal profile – zones and features*

Beaches

A beach is an area of loose, deposited material along the coast, between high and low tide. It may extend above high tide as a storm beach and below low tide in the near shore area. It is a very varied landform, shaped daily by the tides and breakers. The composition and slope of a beach is strongly influenced by the geological structure of the coast. The foreshore (Figure 9.5) is the area on land between high and low tide. The backshore is beyond the highest water level point. The berm marks the boundary between the two. Caribbean beaches vary from the gently sloping, fine white powder of Bahamian and Barbadian coral beaches to the steeper, coarser black volcanic beaches of St. Lucia and Grenada. In between these extremes are brown sand (quartz) beaches of eastern Barbados and eastern Trinidad.

Beaches may be composed of materials such as rocks, sand, gravel, pebbles or cobblestones. Pebbles and small cobblestones form shingle beaches. Some beaches may also comprise shell and coralline algae fragments. The material depends on the nature of the rock type and source of materials.

On some coasts, sand dunes may develop in the back shore. This is the result of wind transporting the dry beach sand inland.

Spits

A spit is a narrow accumulation of sand extending into a bay. It lies just at sea level and is exposed only at low tide. It is connected to land at one end. It is the result of a change in the trend of the coast so that longshore drift continues outwards across the bay. The end of the spit is often recurved towards the mainland by wave refraction. Sometimes the spit may extend to an offshore island to form a tombolo. The Palisadoes, Jamaica, and Scott's Head, Dominica are examples of tombolos.

Bars

A bar is a more general term used to describe coarse, sediment deposits in ocean or stream that shallows water. A spit is a type of bar and it may join across a bay to form a bay-mouth bar, which occurs offshore.

Bars may also form barrier beaches and barrier islands. They are thought to be formed offshore but gradually migrate landward. They are separated from the mainland by a lagoon. These are typified by those off the east coast of the USA. Smaller examples can be seen on the east coast of Trinidad at Matura Bay and Manzanilla Bay, where 16km of barrier beach blocks the Nariva Swamp at the mouth of the Nariva River.

Cuspate bar/foreland

Where bars build out from opposite directions enclosing a triangular lagoon, they form a cuspate bar. Over long periods of time and migration of the bars, a low triangular area of land is formed called a cuspate foreland (Figure 9.6). It consists of ridges of sands separated from each other by belts of marshy ground called swales.

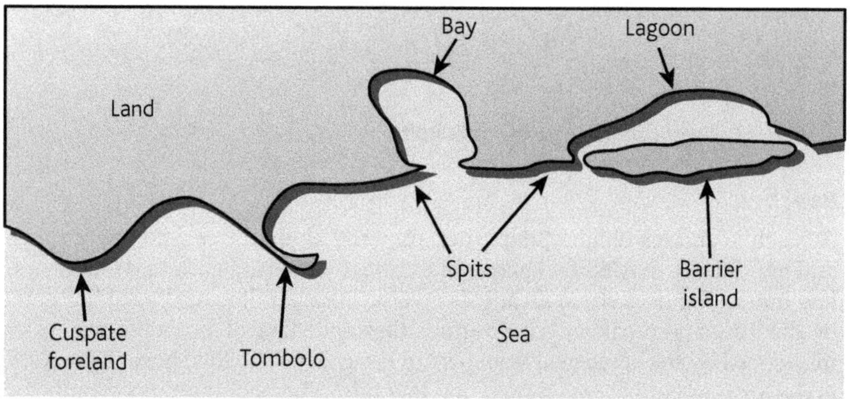

Figure 9.6 *Coastal features*

> **Activity 9.2**
>
> Explain why erosion and deposition tend to be dominant on different parts of a coastline.

Feedback

1 Erosion and deposition will dominate different sections of a coast depending on the interaction between wave energy and geological structure. High energy waves will generally result in erosion as they break powerfully on the shore. They often plunge to shore and drag material out in the backwash. Deposition will occur where there is low energy, weak waves, gently spilling to shore. These conditions result in material being left on the shore by the swash. Wave refraction will also result in erosion of headlands and deposition in bays.

However the geological structure and rock type of the coastline will influence the outcome. Where there are rocks of low resistance, rapid wave action and weathering may produce a beach with a gentle profile; while adjacent areas of greater resistance remain standing as cliffs. This effect is shown on the north east coast of Barbados where the junction between the limestone cliffs and sandstone beaches are formed under similar wave action.

Geological Structure and Coastlines

The nature of the land which meets the waves is of critical importance to the outcome of the land/sea interaction. In both plan and profile, coastal landforms are the result of the geological structure of the shore.

The arrangement of the rocks along the coast is an important determinant of the location of bays and headlands. Generally, bays are created in weaker rocks which erode more rapidly, while resistant rocks protrude as headlands. This is noticeable on the east coast of Trinidad where the weak clays form bays (Matura Bay) and the resistant sandstones form a small cliffed headland (Coco Point). Similarly the more resistant limestone cliffs of St Lucy and St Phillip, Barbados, may be compared with the long beaches on the weaker clays and sandstones of St Andrew and St Joseph. In both islands the potentially destructive wave action on the east coast is influenced by the geological structure.

On the volcanic islands, geological structure also influences the coasts. Where the waves are able to breach a caldera, large circular bays are formed, such as Soufriere Bay, west coast of St Lucia. Lava flows also resulted in the indented coasts of southern Grenada.

The relief of the land, its trend, dip/strike angles, faults and joints all play their part. The north-south faulted valleys of the Northern Range of fold mountains in Trinidad have produced a much embayed north coast. The steep offshore gradient of the north coast of Trinidad stands in sharp contrast to the shallow waters of the west coast.

The characteristics of the rock influence the detailed cross section of the coastal landforms. The bedding planes and differential features are shown on the cliff face whether pitted, smooth or layered. Rates of wave erosion and cliff retreat are affected by the dip of the rocks seawards (faster rates as gravity assists rock fall) or landwards (slower rates).

In addition, sub aerial processes are all affected by the geological structure of the coast. Processes of weathering and mass wasting are controlled by the rock structure and type. Clays are susceptible to slumping and this would aid in the retreat of the cliff once it is formed. More resistant igneous material may weather more slowly maintaining a steep cliff face as it retreats.

Human Factors and Coastal Landforms

The human impact on coastal areas is sometimes so great as to completely mask the natural processes. With more than 6 billion of the world's people living within 60km of the earth's 500 000km of coast, it is rare to find a stretch of coastline unaffected by human activity. Human beings have been attracted to coastal areas particularly near the mouths of rivers for fresh water, fertile soils, ease of transportation and flat land for construction. In the Caribbean tourist developments have had a strong impact on many coastlines. The attractiveness of locations overrides considerations of the threat of flooding and other hazards.

From old Dutch dykes to Guyanese sea walls and Dubai's luxury district, human activity has attempted to harness and restrict the coastal processes. Cities such as New Orleans, USA and Georgetown, Guyana have been built below sea level with levees and seawalls built in attempts to control the sea. The entire country of Bangladesh and large portions of Myanmar (Burma) occupy low-lying, deltaic land subject to annual flooding.

The coastline of Barbados has been so altered by land reclamation, groyne construction, in-filling of swamps and coast rehabilitation schemes, that it is difficult to investigate coastal processes without examining the history of human settlement. Wave refraction, longshore drift and even the beach/cliff profiles have been significantly altered.

Human activities have impacted coastal areas in many ways.

- The infilling and reclamation of coastal mangrove swamps and the destruction of coral reefs both leave the coastline unprotected from erosion. Port and urban development irreversibly change the coastal flows and processes. This is evident in West Kingston, Jamaica, New Port development and in West Port of Spain expansion. Mangrove swamps were drained, and waters dredged to accommodate large ships. The flows and processes were greatly changed and replaced on these sheltered coastlines.
- Removal of beach sand leaves the coast vulnerable to wave attack. The sand is no longer present to dissipate wave energy. This allows larger and steeper waves to erode the back shore.
- Engineered structures such as jetties, groynes, breakwater and seawalls, block longshore drift, producing an accumulation of sand on one side. On the other side erosion is increased as these areas lose their supply of sand. This is evident along the west coast of Barbados where long beach stretches alternate with eroded areas.
- Erection of sea walls and breakwaters attempt unsuccessfully to restrain the sea. This often leads to pile up of wave energy eventually breaching the defences.
- Tourism development results in hotels being built below high water marks with the need to build sea defences. Other tourism developments alter the coast and its processes entirely. For example, Port St. Charles, Barbados, created a deep lagoon for yachts with its sea outlet interrupting longshore drift. It also cut off and blocked one of the few remaining wetlands.
- Global warming is expected to result in changes in sea level. This drowning of coastal areas will impact profoundly on coastal processes and landforms.

Influence of Sea Level Changes

There are two types of sea level change. Isostatic changes are localized. The term refers to a relative change in sea level caused by the loading and unloading of ice. They are therefore changes in the level of the land itself. The formation of large sheets of ice exerted overburden pressure causing the land to subside. The melting of the ice caused the release of ice overburden and the rebounding of the land. In Britain, isostatic readjustments after the last Ice Age are still taking place. Northern

Britain is uplifting at a rate of 2mm y^{-1} while southern Britain is subsiding at a rate of roughly 2mm y^{-1}.

Eustatic changes are global and there are several causes. The first is climatic change. In cool periods, more water is locked up in ice sheets on land and global sea level falls. Water levels in the sea fall relative to the land. In warmer periods the ice melts and the ocean basins are refilled. Sea levels rise relative to the land. The increase in the temperature of the earth has another effect – the thermal expansion of sea water causes sea levels to rise. Global sea levels are also affected by tectonic activity.

A fall in the sea level relative to the land is a negative change. A fall in the land relative to the sea is a positive change.

Landforms caused by negative changes/Emergent coastlines

Coast lines which are above present sea level are easily identified by the landforms of wave action occurring far inland. Raised beaches formed by elevated wave cut platforms can sometimes show the original cliff and cave. Relict notches, stacks, arches and coral reefs may exist high above sea level. These features can be seen in Barbados western terraces. These limestone terraces have stacks and caves amid the sugar cane fields. Ungraded rivers which enter the sea or have waterfalls near their mouths also attest to falling sea level.

Landforms caused by positive changes/submerged coastlines

When the level of the land falls the coasts are drowned. Such coasts may be identified by the shape of the coastline.

Rias result from the flooding of the lowest part of the course to the river and the floodplains. They tend to be winding with gentle valley sides reflecting the nature of the original valley. Many rias are found along the south coast of England – Portsmouth Harbour, the estuary of the Exe River. The Chesapeake Bay is an example on the east coast of USA.

In some cases, glaciers eroded below sea level. When these deep glacial valleys were flooded, fjords were formed as in southern Chile and British Columbia. They are u-shaped with steep valley sides and a shallow section at the seaward end. Its course is straighter than that of rias.

If the valleys are parallel to the coast a Dalmatian coast of long islands and elongate bays is formed.

Formation and Distribution of Coral Reefs

Formation

Coral reefs are colonies of marine organisms sometimes referred to as the 'rainforest' of the sea for its biodiversity. The main reef-building species of coral consists of individual tube shaped polyps, with tentacles for catching food. They support another organism in their tissue which photosynthesizes its food providing oxygen. They reproduce on one night per year. The coral polyps secrete calcium carbonate which slowly builds up the 'reef'. The live coral are found in a very thin layer at the top of the reef.

They are found between 30 degrees north and south of the Equator, where tropical waters are present. Coral reefs need very specific conditions to grow.

- Warm seawater between 25° and 27°C. Colder water retards the growth and warmer water bleaches the corals eventually killing them.
- Sunlight is necessary for photosynthesis. The water must be clear for sunlight to penetrate. Sediment-filled water, such as are found near the mouth of rivers, prevents sunlight from entering. Sediments also choke the corals.
- Corals grow best at a depth of 20–40m. This allows for the penetration of sunlight. The depth of water also affects temperature as temperature decreases with depth of water. Deep water, more than 100m with limited light would limit coral growth.
- Normal seawater salinity along with gentle movement is best for the coral. At the mouth of rivers, where water is fresh, conditions are unfavourable.

Types of coral reefs

Reef building corals require a place of attachment to establish. They colonise the slopes of landmasses or islands. There are three basic types of coral reefs: fringing, barrier and atolls.

In the Caribbean, most islands have some coral reefs offshore: from the lone barrier reef of Belize through the fringing reefs of Jamaica, the Eastern Caribbean volcanic islands, Barbados and Tobago. The sediment-rich waters of Trinidad and Guyana (fluvial coastal environments) prevent coral reef growth except for Salibia Bay off the north east coast of Trinidad.

Fringing reefs are the most common types found within the Caribbean. They occur very close to the shore and are separated from it by shallow water. As waves break on the reef, they protect the coast from wave erosion. The waves reaching the shore are low energy spilling breakers. Buccoo Reef off south western Tobago is separated from land by water only 3m deep. Jamaica has almost continuous fringing reef along the western, northern and eastern coastline.

Barrier reefs form farther from the shoreline and are separated from land by deeper water. The Australian barrier reef is one of the largest barrier reefs in the world. The barrier reef located off the coast of Belize is the second largest in the world at 200km long. The reef is also wider and larger than fringing reefs.

Atolls are small circular reefs enclosing shallow lagoons, but without any land inside. There are few true atolls in the Caribbean similar to those of the Pacific Ocean. The atoll-like features off Belize are thought to have been formed in a different way.

Theories of coral reef formation

Uplift and subsidence theory

Charles Darwin argued that the growth of coral reefs began with the building of fringing reefs around an extinct volcanic island. The volcano sinks lower in the sea and the corals grow upwards to ensure continued access to sunlight. The outer side of the reef grows faster because ocean currents supply plankton. The inner side is submerged. In the submerged area between the island and the coral rock, a lagoon forms. A barrier reef develops. With further subsidence, the summit of the central island sinks and an atoll is formed.

Dana agreed with the subsidence theory although there were small areas of disagreement. For example, sections of the atolls may be wider than the rest and Darwin believed that the differences were caused by the topography of the island's coastlines. Gently sloping coastlines, he said, produced wider reefs; steep coastlines would produce narrow reefs. Dana, however, felt that the shape was the result of the actions of winds and waves and that wider areas were those that had a more plentiful supply of nutrients.

Glacial control theory of Daly

The uplift and subsidence theory could not explain why the floor of the lagoons were so flat and of uniform depth. Daly's theory also begins with a fringing reef. During the Ice Age, he argued, sea levels were lowered. The part of the reef that remained above water was eroded by sub aerial forces. At the end of the Ice Age sea level rose and corals began to grow once more. This time they grew on a flat surface. This occurred repeatedly leading to the formation of several platforms and terraces.

The theories of these three, taken together, explain the characteristics of atolls today. Scientists refer to the Darwin/Dana/Daly or 3D theory of reef formation.

Threats to coral reefs

Coral reefs are dying because of human activity as well as from natural causes. They are dying from:

- Harmful fishing practices such as the use of dynamite and cyanide to catch fish. These destroy both the habitat for fish and the biodiversity that is necessary for the maintenance of a healthy reef. Overfishing removes species such as the sea snails which keep predators that eat corals under control. There is selective removal of species for the ornamental trade.
- Poor land use practices which produce run off and sediments that smother the corals or produce diseases that harm the corals.
- The destruction of mangrove forests that serve as nurseries for reef species and reduce the run off of pollutants. Mangrove forests are removed to make way for the growing of rice, palm oil and the production of pond shrimp for the world market.
- Corals are also harvested for the tourist trade. Boats used in the tourist industry are also harmful to corals.

In addition:

- The rise in sea temperatures is associated with coral bleaching and the death of corals.

Activity 9.3

1. Explain how fringing reefs are formed.
2. Name and locate one barrier reef in the Caribbean.
3. With reference to specific examples, explain how two different types of human activity threaten coral reefs.

Feedback

1. Fringing reefs are formed when new geological structures provide platforms for coral polyps to colonise. In the clear shallow waters in the near shore area the polyps deposit calcium carbonate. The living opening continues to grow upwards. The platform subsides as the coral grows to maintain the shallow, clear water conditions required.
2. The only barrier reef found in the Caribbean is the Belize Barrier Reef. It lies offshore of Belize in the western Caribbean.
3. Coral reefs are threatened by increasing tourism activities. From the yachts which cut off the coral heads with their anchors, to the cruise ship waste and tourist groups walking on the coral, tourism has a severe impact on the fringing reefs. On the coast of Barbados, Jamaica and Puerto Rico, even with stringent legislation, many of the reefs are struggling to maintain their

growth. Intense use of the coast for water sports and by many tourism water crafts disturbs the natural conditions for coral growth.

The second major threat is in pollution of the coastal water. The addition of various materials changes the chemical composition of the water and kills the reefs. For example, in Jamaica, agricultural run-off nourishes the algae which smother the coral reefs. Similarly pollution by sewage also upsets the ecological balance allowing the excess growth of bacteria and algae. Deforestation of the land increases the sediment washing into the sea, blocking the sunlight. All these pollutants create environmental stress for the coral reefs.

Key Points

- Wave action consists of flows such as longshore drift, wave refraction and currents, as well as processes of erosion and deposition.
- Wave action results in many landforms such as cliffs and wave-cut platforms as well as beaches, spits and bars.
- Coral reefs produce distinctive coastal environments.
- Geological structure and human activity have a strong influence on coastal environments.
- Changes in sea level impacts on coastal landforms.

Conclusion

Coastal environments are important resources for most Caribbean territories. The processes of wave action shape cliffs and beaches. Processes such as weathering are active on coastal landforms. Factors such as geological structure and human activity also affect the shape and form of coastal landforms. Management of coastal resources is important to the future development of the region.

End Test

For questions 1–5 choose the letter corresponding to the correct answer.

1 Wave height is the vertical distance between:

a wave crest and trough

b ocean floor and crest

c two wave crests

d sea floor and wave top.

2 The fetch of a wave is the:

a length from crest to crest in open water

b height from crest to trough near the shore

c maximum distance of unbroken wind flow in open water

d maximum distance of swash on the shore.

3 Longshore drift combines waves approaching the shore at an oblique angle and returning to the sea at:

a an oblique angle

b a right angle

c an acute angle

d parallel to coast.

4 Tidal variations are the result of :

a spring and summer tides

b tidal waves or tsunamis

c gravitational forces of sun and moon

d inflow in river mouths.

5 The breakdown of rocks into finer particles is accomplished by the process of:

a corrosion

b hydraulic action

c corrosion

d attrition.

6 Indicate whether the following are true or false:

a Tectonic movements have no effect on coastlines.

b Waves of oscillation move water in a circular motion.

c Wave refraction results in straightening of the wave crest.

d Urbanization does not affect coastal processes.

e Sewage outflows are a major threat to coral reefs.

7 Draw a well-labelled diagram to show the features of a typical beach profile.

a Explain how sea cliffs are shaped by wave action and other non-marine processes.

b Explain two ways in which human activity impacts on coastal processes.

End Test Feedback

1 a **2** c **3** b **4** c **5** d

6 a False b True c False

d False e True

7 Diagram

a Sea cliffs can be formed directly by tectonic processes. Emergence of steep surfaces, for example raised coral reefs, or steep volcanic slopes can result in sea cliffs. Once emerged, wave action can produce cliffs through its erosion of the land at sea level. This erosion will eventually lead to the formation of notches and caves. The resulting rock overhang of the roof of the cave may collapse to form a steep faced cliff. Repeated erosion and collapse may form a beach covered wave platform. The cliff retreats in this way until the wave energy is dissipated on the wide wave cut platform. The cliff is then no longer active.

Sub aerial processes and the geology of the coast then influence the detailed form of the cliff. It will retreat rapidly on such weak rocks as clay where slumping and sheet wash will aid wave erosion. Limestone cliffs may show the pitting of chemical weathering while maintaining a steep angle. Differential weathering of joints and cracks also aid in the breakup of the overhanging rocks. Frost shattering in temperate areas would increase the rate of cliff retreat.

b Human activity has a very strong influence on coastal processes.

One way is the building of groynes and breakwaters. These interrupt long shore drift. It is accelerated on one side and depleted on the other. Instead of a smooth movement along the coast in the direction of the prevailing wind, there are area of great accumulation and erosion in adjacent areas.

A second way is the reclamation of land for the sea. This pushing the waves further seaward often increasing their erosive effect since the water is now confined to a shorter run up to the shore. This increases wave height and often results in plunging breakers.

10 Processes and Landforms in Limestone Regions

General objective

On completion of this module you should understand the geomorphic processes in limestone areas.

Specific objectives

You should be able to:

- explain the main concepts, flows and processes associated with limestone environments
- explain limestone processes influencing the development of landforms
- analyze the physical, biotic and geological factors which affect processes operating in limestone areas
- explain the effect of human and physical interaction within the limestone areas over time.

Introduction

Limestone areas form distinct landscapes controlled by the chemical composition and structure of the rock. The landscape can be recognized for the dominance of chemical weathering, the absence of surface drainage and unique limestone landforms, such as depressions and residual hills. Landscape formed by the solution of limestone is referred to as 'karst', the word used originally in former Yugoslavia. Many Caribbean countries, such as Jamaica, Puerto Rico, Belize and Barbados, have large areas of limestone. The Cockpit country of western Jamaica is one of the areas in the Caribbean where karst landscape is well developed.

Content

- Characteristics of limestone as a rock.
- Chemical weathering processes and limestone.
- Characteristics and development of limestone landscapes.
- Factors influencing processes in limestone areas.

Characteristics of Limestone as a Rock

Limestone consists of more than 50 per cent calcium carbonate and makes up about 10 per cent of all sedimentary rocks. It is found on most continents and under different climatic conditions such as, British Carboniferous limestone, limestone of Guilin, China and south west USA. There are many types of limestone which may be classified by texture, mineral content (for example, fine oolites), origin (for example, coral) and geological age (for example, Carboniferous). Oolites, magnesium-rich dolomites, soft, porous chalk and coarse, coral limestone rocks are all classified as limestone although they vary greatly in colour, texture and chemical composition.

Characteristics:

- *Chemical composition*: Calcium carbonate ($CaCO_3$) is reactive with acids. This makes it susceptive to chemical weathering and other water-related processes. In Jamaica, White Limestone Group is very pure while the Yellow Limestone contains other minerals.
- *Hardness*: Although this varies with age and origin of limestone, they are generally considered soft rocks (3–4 on Mohs 1 to 10 scale of hardness). Differential rates of weathering with surrounding rocks may result in some being prominent as escarpments and others worn flat.
- *Texture*: Coral limestone, as found in Barbados, has a coarse texture reflecting its coral origin. Others such as oolites, composed of spherical grains called ooids found in shallow marine waters in Bahamas, have finer textures. Some limestone may consist of mineral crystals, that is, they are crystalline as in Jamaica.
- *Structure*: Most limestone have many planes of weakness – bedding planes, joints, cracks and faults. The rocks are pervious rather than porous. Weathering is accelerated along these joints, widening them and allowing water deep into the rock.

Chemical Weathering Processes and Limestone

Limestone rocks are particularly susceptible to solution by carbonation. In this process, water which is a weak carbonic acid, reacts with the calcium carbonate to produce calcium bicarbonate which is soluble and easily removed.

The reaction is presented as the chemical equation:

$CaCO_3 + CO_2 + H_2O \longleftrightarrow Ca^{2+} + 2HCO_3^-$

Limestone dissolved limestone

In caves and other places where the calcium-rich waters are evaporated the equation is reversed and limestone is formed inorganically. The calcium is re-precipitated as travertine in the form of stalactites and stalagmites.

The term 'karstification' is used to describe the processes by which rocks, such as limestone, are subjected to the solution process.

Activity 10.1

1 Name one rock rich in calcium carbonate, other than limestone, which is susceptible to chemical weathering by carbonation.

2 Describe the process by which limestone is chemically weathered.

3 Describe two significant characteristics of limestone.

4 Explain how these two characteristics make it susceptible to karstification.

Feedback

1 Dolomite

2 Carbonation really consists of two processes: the first one is the absorption of carbon dioxide in the air and soil into water. This is usually given by the formula

$CO_2 + H_2O \longleftrightarrow H_2CO_3$

This forms a weak carbonic acid.

The second process occurs when this acidic water reacts with limestone changing it to the soluble form calcium bicarbonate. It is then easily washed away. This process is presented as

$CaCO_3 + CO_2 + H_2O \longleftrightarrow Ca^{2+} + 2HCO_3^-$

Limestone dissolved limestone.

3 Answers could include:

it is susceptive to chemical weathering.

it has many planes of weakness.

it is permeable and pervious.

4 Limestone is particularly susceptible to karstification because of its chemical composition and structure. By definition it is composed of at least 50 per cent calcium carbonate. This is acted upon by carbonation which results in solution landforms known as 'karst'.

Secondly the presence of joints and cracks allows the process to occur deep below the surface. The cracks and joints are widened by solution resulting in the depression and hills typical of karst. Moisture seeps between the spaces and solution occurs within the rock. This creates the underground features which are typical of karst.

Factors Affecting Chemical Processes in Limestone

Many factors are responsible for the occurrence and rate of chemical weathering processes.

These include:

- *The purity of the limestone*. Karst is best developed in very pure limestone, for example, in the 98 per cent pure Jamaica White Limestone. Less pure limestone slows down the carbonation process and creates residues.
- *Rock structure and lithology*. The permeability of the limestone allows weathering deep underground. Its coherence allows it to support underground cavities and towers.
- *Water*. Water is a major control on the dissolution of limestone. Surface and underground flow and ground water all contribute to the erosion of limestone. Chemical processes require water to be present.
- *The amount of dissolved CO_2*. The rate of absorption of carbon dioxide by water increases with decreasing temperatures. However the rate of carbonation is increased in warm temperatures. In addition, humic acids, sulfuric and nitric acids (acid rain) will contribute to the dissolution of limestone.
- *Climate*. The combination of temperature and precipitation regimes has a great influence on the occurrence and rate of chemical weathering. Dry, cold climates slow chemical weathering even though the concentration of carbon dioxide may be high. The lack of water would also retard the dissolution of limestone. Karst features are best developed in humid areas.
- *Vegetation*. Vegetation cover aids in the process of chemical weathering by the production of humic acid from decayed organic matter.
- *Time*. Old limestone which have been exposed to chemical weathering for millions of years show a greater complexity of drainage and both surface and subsurface landforms. The processes of dissolution and collapse and disrupted drainage result in the typical 'karst' landforms.

These factors influence the outcome of the solution process resulting in the pitting and depressions of limestone surfaces. The general absence of mass wasting processes and surface run-off also adds to the particular nature of these environments.

Characteristics and Development of Limestone Landscapes

It is difficult to generalize about limestone landforms. More than any other geomorphic environment, the specific characteristics of the local geology control the nature of the landscape. Karst landforms vary greatly with time, chemical composition and the structure of the rock. Only in a very basic sense can these landscapes be said to share:

1. etched, pitted surfaces with thin soils
2. circular basins and depressions
3. residual hills
4. dry valleys
5. underground caves and caverns
6. disrupted surface drainage and underground drainage.

A further difficulty in generalizing about karst landscapes arises from the many local names for similar features.

Common features of limestone environments:

1 Karren

This is the collective name given to small scale grooves and etching commonly found on limestone pavements. This includes clints (small ridges) and grykes (grooves). Some grykes in Yorkshire are up to 2m deep and 0.5m wide. There are very thin soils on limestone rocks. The dominance of chemical weathering leaves a thin residual layer on the surface with little soil development.

2 Depressions

- *Dolines/sinkholes*. These features may vary greatly in shape and size. They may be cylindrical, conical or bowl shaped and can be a few metres or hundreds of metres in width and depth. They generally drain underground. They are subdivided into solution formed as solution progresses along a joint, subsidence and collapsed doline.
- *Uvalas are larger depressions with irregular floors*. They are thought to represent the collapse of sides or coalescence of adjacent sinkholes.
- *Poljes*. Polje is sometimes referred to as a karst valley. It is broad and flat-floored often associated with faulting. They are well-developed in Jamaican limestone.

3 Residual cones and towers

The areas which remain after solution form cones or towers giving karst landscape a hummocky topography. The evenness of the height of the hills is very noticeable as it represents the height of the former surface.

- *Cone karst*. These steep-sided hills rise from surrounding depressions. Cockpit karst is one type of cone karst.
- *Tower karst*. These are tall, steep sided protrusions. They are thought to originate as cones but are steepened by water table undercutting. Some examples of tower karst are found in Puerto Rico, Cuba and southern China.

4 Dry valleys

Steep-sided valleys without rivers are common in limestone areas. They represent valleys which may have contained water in wetter times or before the stream went underground. The collapse of a series of underground caverns will form a gorge.

5 Caves and caverns

The water passing underground dissolves the limestone internally along joints and cracks forming underground caves, caverns and passages. These interior open spaces usually have an entrance, one or more chambers, passages and termination. These holes migrate upwards as the roof collapse and debris collects on the floor. Eventually they open to the sky as karst windows. They usually contain speleothem, which are the many forms of re-deposited travertine (limestone) in them – stalagmites, stalagtites, columns and flowstones.

6 Disrupted surface drainage and underground drainage.

Areas of limestone are often noted for an absence of an integrated surface drainage pattern. Valleys may be present without any water in them (dry valleys). Many areas show an absence of surface drainage, or may have a few seasonal streams. In other areas, streams may disappear and reappear in broken strands. This complex drainage pattern is the result of the disappearance of streams down swallow holes, flowing underground and then resurfacing (resurgence) when they meet impermeable layers or the water table.

In addition, a rising water table during rainy periods may flood some depressions. Residual clay deposits are redistributed by this flooding and block the floor of the depression.

Feedback

1 a 2 c 3 b

Activity 10.2

1 Karst landscapes are associated with:

- a absence of surface water
- b gently sloping even plains
- c lack of aquifers
- d high fold mountains.

2 When neighbouring sink holes coalesce they form a:

- a doline
- b swallow hole
- c uvala
- d cave.

3 Travertine is the name given to:

- a dissolved limestone
- b re-precipitated limestone
- c caves and caverns
- d swallow holes.

Caribbean karst

Karst in the Caribbean is associated with limestone dominated by roof collapse of underground passages. Limestone of varied ages is found in many Caribbean countries, such as Jamaica, Puerto Rico, Barbados and Belize. The hot, wet tropical marine climate has given rise to the distinct karst features in many territories.

In Jamaica, the karst of the Cockpit Country is the type location for this tropical karst. The White Limestone formation covers 70 per cent of the island and is of very high purity. The limestone plateau rises to 915m above sea level. It is of great thickness and well faulted. It has a grid of joints east-west and north to south.

This area receives more than 2000mm of rainfall per year, which has resulted in karstification of the limestone. This cockpit karst consist of steep-sided closed depressions (cockpits) separated by conical hills. The cockpits are often star-shaped. The cockpits are usually drained by sinkholes. During rainy periods they may be flooded by rising water tables. This water may help to undercut and steepen the base of the hills or mogotes.

Jamaica also has many poljes or interior karst valleys. They may be elongated with steep sides. They may be occupied by seasonal streams.

Key Points

- Limestone environments are distinctive as a result of the special characteristics of the rock.
- Its chemical composition makes solution a dominant process.
- Many factors such as geological structure and water influence the exact nature of the landscape.
- Karst scenery is marked by little surface drainage, underground features, depressions and circular hills.

Conclusion

The characteristic chemical composition and structure of limestone make it susceptible to chemical weathering processes. The dissolution of the limestone results in a hummocky landscape of alternating depressions and hills. The absence/disruption of surface drainage is also a significant feature of karst landscape. Underground caves are often present. Although karst occurs under many environmental conditions, it is best developed in warm, humid climates such as in the Caribbean.

End Test

1. What is meant by karst topography?
2. Give two reasons why karst topography is not well developed in Arctic regions.
3. Explain the formation of two surface features formed by the dissolution of limestone.

End Test Feedback

1. Karst is the name given to the distinctive hummocky relief and drainage found in limestone areas. It is the result of solution of limestone by surface water acting as a weak carbonic acid.
2. Karst is not well-developed in Arctic regions because it is too dry for solution to occur. There is little running water in these cold regions. In addition it is too cold for biotic activity to produce humic acid which aids the development of karst. The surface of Arctic regions is frozen and water cannot go underground.
3. Depressions and residual cones are formed from the solution of limestone. There are many different types of depressions or circular basins. Some are formed from enlarged joints. Water passing through the joint dissolves the limestone along this weakness until it forms a depression. In special cases where joints are at right angles there may be star-shaped large depressions (Jamaican mogotes). Other depressions are formed when two or more sink-holes coalesce. The sink holes are enlarged until they form one large opening called an uvala. Some depressions may be the result of the collapse of underground features. The slow solution underground may so undercut the ceiling of the cave that it collapses on itself. This forms a depression on the surface.

   Residual hills have formed cones or towers in the karst landscape. They are noted for the evenness of height and very steep sides. They alternate with depressions to give a hummocky profile. These hills represent what is left of the original surface. Solution has removed the intervening areas.

11 Natural Events, Hazards and Disasters – Flooding

General objectives

On completion of this module you should:

- understand and appreciate the differences between natural events, natural hazards and disasters
- appreciate the relationship between the natural and human environments in causing hazardous floods.

Specific objectives

You should be able to:

- distinguish between a natural event, a natural hazard and a disaster
- describe the different types of hazards
- explain the causes of flooding.

Introduction

There are many extreme natural events such as the volcanic eruptions in Montserrat, severe earthquakes such as that which devastated Haiti in 2010 and the floods that submerged parts of northern England in 2009 and the island of Madeira in 2010. These physical events help to shape the landscape and are regarded as natural phenomena. Human beings occupy most of the earth's surface and these natural occurrences become hazards when they threaten human life and property. When a hazardous event results in a very large number of casualties or extensive damage to property it becomes a natural disaster. Natural hazards and disasters are not the destruction caused by the unavoidable forces of nature. Disaster marks the interface between a physical event and a very vulnerable population.

Populations are vulnerable not only because they occupy areas that may be in the path of a hazardous event. Disaster vulnerability is expressed in the physical, social and economic life of the country. The following criteria are used to measure the vulnerability of countries:

- Exposure (for example, population growth and density, GDP, poverty).
- Socio-economic fragility (for example, dependency ratio, unemployment, social disparity, debt servicing, inflation).
- Lack of social resilience (for example, HDI, GDI, expenditure on welfare, pensions).

The response of human beings to hazards can vary greatly and may depend on the level of development, the culture, individual perception of the event as well as the adjustments it is possible for them to make. Adjustment is the response individuals make to hazards before, during and after the event. Sometimes collective action is called for and the response is on a community wide basis. There are some types of responses that only governments can make to mitigate the impact of a hazard/disaster.

This module will focus on distinguishing between natural events, natural hazards and disasters and types of types of hazards. It will then look at flooding as a natural event. Flooding is one of the most common types of natural events on the earth's surface. Inundation of the land by water is inevitable on the 'blue planet'. The land adjacent to rivers, lakes or oceans may become covered with water for many reasons. Flooding may be the result of precipitation events (Module 7 – Hydrological Processes), drainage basin characteristics (Module 8 – Fluvial Processes and Landforms), sea level changes (Module 9 – Coastal Processes and Landforms) and the influence of human activities.

However, flooding is hazardous because a high percentage of the world's population lives in these potentially 'wet' locations. The attraction of fresh water, fertile soils and water transport has resulted in very high population densities in river valleys, lakeside and coastal locations. In addition, by replacing permeable soil by impermeable concrete surfaces, and in damming and canalizing rivers, human activity has aggravated the hazard of flooding.

Flooding demonstrates very clearly the complex relationship between the natural and human environments. Human beings are dependent on the natural environment for many of their needs but at the same time they have a profound influence on it.

Content

- Concepts of natural events, hazards and disasters.
- Types of hazards: technological, tectonic, climatic, geomorphological.
- Types of floods: riverine, coastal and estuarine.
- Causes of floods:
  types of precipitation events
  drainage basin characteristics
  sea level changes
  influence of human activity.

Natural Events, Hazards and Disasters

The natural events are the physical processes that shape the earth. They are naturally occurring phenomena such as the eruptions of volcanoes, hurricanes and landslides. These are not new processes but have been at work throughout the earth's history. They are now becoming more hazardous because they are affecting more people and more property. Technological developments are also creating chemical leaks and nuclear accidents.

Disasters are created when the impact of the hazard creates widespread destruction and distress. One definition suggests that disasters represent 'a situation which overwhelms local capacity necessitating external assistance'.

Activity 11.1

Indicate whether the following statements are True or False:

1 All extreme natural events are hazards.

2 All hazards are caused by natural events.

3 Events are only hazards when they impact on human life and property.

4 Technological hazards are those associated only with computers.

5 Not all hazards are disasters.

6 Drought is not a true hazard as it is slow-forming.

7 Nuclear leaks are not natural hazards.

8 Technological hazards are the result of human activity.

Feedback

1	False	5	True
2	False	6	False
3	True	7	True
4	False	8	True

Types of Hazards

Hazards may be classified according to the causal process.

- *Technological hazards* are those caused by human activity, for example, collapse of construction machinery and mines, acid rain and nuclear leaks. For example, the collapse of construction cranes in New York in 2008 was a technological hazard. Larger disasters like the Chernobyl nuclear leak in Russia in 1986 and the chemical leak in Bhopal, India were both technological hazards caused by human activity.

- *Natural hazards* are the result of physical processes.

  Climatic conditions, such as hurricanes, high temperatures and low rainfall can be hazardous to human life and property. Hurricane Ivan in 2004 created disastrous conditions in the Caribbean. Heat waves in France resulted in deaths of many elderly persons in 2007; while drought conditions have occurred in Cuba and parts of Africa in successive years.

  Tectonic events of volcanic eruptions and earthquakes cause dramatic and extreme natural hazards. The tsunami of Thailand 2006, the Chinese earthquake of 2008, the Haitian earthquake in 2010 and volcanic eruptions in Montserrat after 1996 have had disastrous impacts on human life and property. Hundreds of thousands of lives have been lost and there has been billions of dollars worth of damage to buildings and agriculture.

  Geomorphologic processes such as landslides and flooding can also be hazardous. The classic landslide of Alberta, Canada in 1903, involved an estimated 27 million cu m falling 900m and burying the town of Frank. River flooding takes many lives annually across the world, for example, in Bangladesh in the delta of the Brahmaputra River.

However all the hazardous areas of the earth have positive aspects which attract large populations. River floodplains have supported people with fresh water, fertile soils, and water transport for nearly all of human settlement. Erupting volcanoes create fertile soils; while hazardous climatic conditions have such positive benefits in high rainfall.

The Table 11.1 below shows some examples of types of hazards.

Table 11.1

Causal process	Event	Examples of hazardous conditions
Tectonic	Volcanic eruption	Ash falls, lava flows, pyroclastic flows
	Earthquake	Tremors, fissures, tsunamis
Climatic	Hurricane	Heavy rainfall, strong winds, storm surge
Geomorphic	Landslides	Blocked river, unstable slopes, floods
Technological	Nuclear leak	Toxic air and land conditions

Activity 11.2

1 Complete the table below by indicating the causal process responsible for the hazard stated at left.

Hazard	Causal process
Avalanche	
Flood	
Volcano	
Tornado	
Landslide	
Nuclear leak	

2 Explain the difference between a 'natural hazard' and a 'technological hazard'. Give two examples of each in your response.

Feedback

1

Hazard	Causal process
Avalanche	Geomorphic
Flood	Geomorphic
Volcano	Tectonic
Tornado	Climatic
Landslide	Geomorphic
Nuclear leak	Technological

2 A natural hazard is one caused by natural events such as climatic conditions while a technological hazard is one created directly by human activity. Examples of natural hazards are earthquakes and hurricanes; while examples of technological hazards are acid rain and nuclear leaks.

Flooding as a Natural Event-Types of Floods

Flooding occurs when more water remains on the land surface than can be contained in river channels, or removed by surface runoff. Floods may be classified by the area which is flooded as a result of different processes.

- *Riverine* floods are one of the most common types. These occur when the river channel cannot contain the available water and overflows onto adjacent land. Most large rivers flood periodically as a result of seasonal fluctuations in discharge. This creates their large fertile floodplains as they approach the sea. The Mississippi, Ganges, Nile and Hwang-ho Rivers all flood periodically. In the Caribbean, the Caroni River, Trinidad; Black River, Jamaica and the Essequibo River, Guyana, flood especially during the hurricane/rainy season.
- *Flash floods* are local floods of great volume and short duration. They are fast occurring floods in dry channels especially in mountain canyons. They also occur in deserts and in urban areas. Extreme precipitation events such as severe thunderstorms, can deposit large amounts of water on the surface in a short period. The water picks up loose material on dry surfaces and moves rapidly downstream with little warning. These floods are short lived as the water seeps into the surface. For example, in the South west USA in Arizona, flash floods can suddenly create hazardous conditions as dry canyons become raging torrents. The key elements in flash flooding are rainfall intensity and duration.
- *Coastal* flooding occurs under particular conditions. Unusually high tides or storm waves can cause water to cover areas above high tide. For example, storm surges associated with hurricanes can push water onto the land. A storm surge is a wide dome of water pushing onshore near to the eye of the hurricane influenced by both the wind and the low pressure. But water draining from the land can also cause coastal flooding, for example, if the water table is very high; or surface run-off is not being taken away quickly, for example, New Orleans during Hurricane Katrina, and Guyana's annual ITCZ rainfall activity. In addition coasts can be flooded more slowly by sea-level changes, for example, the Maldives.
- *Estuarine flooding*. Adjacent low-lying areas are easily flooded by tidal bores or waves pushing water up the bay. As it is confined by the bay, the waves can reach great heights and flood the land, for example, River Severn and the Bay of Fundy, Canada.

Activity 11.3

1 Distinguish between two different types of floods.

2 Explain why extreme precipitation events are often associated with all types of flooding.

Feedback

1 River floods occur along the course of the river and result when the channel cannot contain the available water. It therefore overflows its banks onto the adjacent land. This flooding can occur far inland. On the other hand, coastal flooding refers to water encroaching on the dry land often from both land and sea. Coastal flooding of the shore is often brackish or even saline and limited in extent inland.

2 Precipitation events are often involved in all types of flooding since this is the main source of water reaching the earth's surface. Intense storms which deliver more water than can be infiltrated or quickly run off in channels will result in flooding of any area. The particular nature of the area of flooding whether riverine, estuarine or coastal will determine the specific type of result.

Causes of Floods

The water in lakes, rivers and the sea, all ebb and flow. Very rarely are water levels static, so it can be expected that floods will occur. They are often predictable natural events.

- *Types of precipitation events*: Flooding can be caused by extreme rainfall events such as hurricanes but also by low intensity events, depending on duration and antecedent soil moisture. More water arrives at the surface than can be absorbed or channelled away. The storm or flood hydrograph is useful in predicting and investigating the relationship between the precipitation event and the river discharge.

  Storm winds can drive coastal water onto the shore as a storm surge. If it happens at high tide, large areas of the coast can be flooded.
- *Sea level changes*: The boundary between ocean and continents has changed over geological time. Depending on the amount of water stored as ice relative to the amount in sea basins, the average sea level can change. In colder geological eras, sea level has been lower than present with larger areas of dry land. In warmer periods as at present, more water is in the ocean basins and sea level is rising. Drowned coastlines and raised coastal features are evidence of this changing boundary. These were considered in Module 9 – Coastal Processes and Landforms. Currently, increased average temperatures of global warming and melting ice caps predict a rise in sea level. This would lead to flooding of heavily populated coastal areas.

  In addition, a rise in sea level raises the base level of the river and increases the risk of flooding as the river deposits its load in order to grade its bed. The lower course will be permanently flooded or drowned.
- *Drainage basin characteristics*: Flooding of the lower course of rivers is a natural and regular phenomenon. As the river approaches the sea with its maximum discharge and flowing above the valley floor, it often exceeds its raised channel and bursts the natural levees to pour sediment-laden water over its 'flood plain'. This was discussed in Module 8 – Fluvial Processes and Landforms.
- *Influence of human activity*: Human activity has a profound effect on the hydrological cycle as discussed in Module 7 – Hydrological Processes. Urbanization covers surfaces with impermeable roads, parking lots, roofs and pavements. This reduces infiltration and increases overland flow. High river discharge easily floods these areas. In addition, the construction of storm drains channels water quickly to the streams. This shortened arrival time often results in flooding.

Key Points

- Natural events, hazards and disasters differ in their effect on human life and property.
- There are many types of hazards with both natural and human causes.
- Flooding is a common natural event which can be caused in many ways.

Conclusion

The relationship between the natural and human environments will always have both positive and negative impacts on human activity. Human beings are completely dependent on the natural environments for basic needs such as water and they have a great impact on the water cycle. Many natural events are hazardous to human activity and some human activity is hazardous both to humans and the environment.

Flooding is a common natural event. The attractions of available water and fertile soils in river valleys have attracted large populations which in turn are vulnerable to seasonal flooding disasters. Although flooding hazards can be lessened by flood management and prevention strategies, the population pressures for space will always put us in harm's way.

End Test

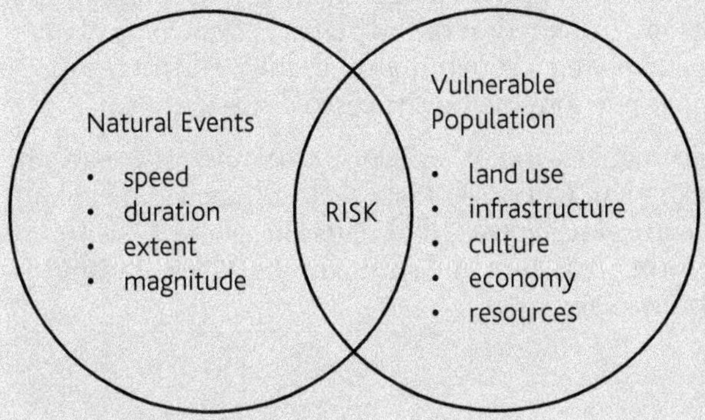

Figure 11.1

With reference to the diagram above, describe the concept of risk.

End Test Feedback

According to the diagram, risk occurs at the intersection of natural events and vulnerable populations. For example, the speed, duration, extent and magnitude of the natural event impacting on the nature of the vulnerable population's infrastructure, land use and so on will determine the hazard. A low magnitude event may present a low hazard risk to a highly prepared vulnerable population.

12 Plate Tectonics

General objectives

On completion of this module you should:

- understand the concept of plate tectonics and associated features
- appreciate the relationship between the natural and human environments.

Specific objectives

You should be able to:

- explain as natural events, the formation and processes involved in the movement of plates
- describe the distribution and characteristics of landforms resulting from plate movements including earthquakes and volcanoes.

Introduction

Plate tectonics explains the processes behind many natural events, such as earthquakes and volcanic eruptions. 'Plate tectonics' describes the concept: 'plate' referring to large slabs of material on the earth's crust and 'tectonic' refers to the forces that produce change in the structure of the earth. The movements and interaction of the plates is known as plate tectonics. This concept explains, though not quite fully, the occurrence and distribution of many phenomena on the earth's surface.

The distribution of earthquakes and other tectonic features is consistent with the assumptions of plate tectonics. Plate tectonics helps to explain the processes which create earthquakes, folding/faulting and volcanic activity. It assists in better prediction and mitigation of these natural events as hazards to human activity.

Content

- Continental drift and plate tectonics.
- Formation, distribution and movements of plates.
- Processes operating at different types of plate margins and hot spots.
- Earthquakes.
- Distribution and characteristics of volcanoes.
- Island arcs and fold mountains.
- Positive impact of volcanic and earthquake activity.
- Value of folded and faulted landscapes.

Continental Drift and Plate Tectonics

The theory of plate tectonics which was developed in the 1960s, confirmed the massive land shifts first suggested by Alfred Wegener (1912) in his *Theory of Continental Drift*. Wegener had proposed that continents were not always in the relative positions they currently occupied. He traced the fossils of plants and animals across continents and found a range of evidence in support of his thesis:

- Identical fossils in rocks in continents widely separated by oceans.
- Rocks, fold belts and mountain belts which would be contiguous if the continents were fitted together.
- Coal and evidence of glaciations in areas whose distribution could not be explained by current climatic conditions.
- The geographical fit of the continents.

He named the original super continent Pangaea with its sea Tethys. He further went on to propose that Pangaea had broken into two continents: Laurasia to the north and Gondwanaland to the south.

Wegener, however, could not explain how continents could move, that is, the mechanism/energy which could fuel these movements. His theory was rejected as ridiculous. Now that plate tectonics has provided the mechanism of convection currents in the molten magma of the mantle, it is widely accepted that continents move and in fact, these movements are measured and predicted.

Earth structure

Only the outermost few kilometers of the earth's interior structure have been directly investigated. The structure of its 6370km radius is mainly inferred from the passage of earthquake waves through it.

At the centre of the earth is the core. It is very hot (2900+ °C) and very dense consisting of iron and nickel minerals, extending for 3478km. The outer core is liquid while the inner is solid.

Around this is the mantle, consisting of olivine rich rocks. The temperature of the mantle increases with depth. Rocks in the upper mantle are cool and brittle enough to break under stress. Rocks in the lower mantle are hot and soft and flow rather than break. Differences in behaviour separate the upper from the lower mantle.

The thin surface layer is called the crust. The thinner high density oceanic crust consists mainly of magnesium silicate rocks. The continental crust, composed of aluminium silicate rocks is thicker and of lower density, and this allows it to 'float' on the mantle.

The crust and the uppermost mantle form the rigid lithosphere. The more mobile lower mantle forms the asthenosphere. The two are separated by the Mohorovicic discontinuity.

Evidence supporting Plate Tectonics

Since Wegener's theory of Continental Drift was rejected, advances in technology and science have accumulated evidence to support plate tectonics.

- 1948 – Maurice Ewing noted a continuous mountain range the whole length of the Atlantic Ocean bed, and that these rocks were volcanic and recent.
- 1950s – Paleomagnetism established that the pattern of magnetic reversal in the rocks on either side of the Atlantic ridge were identical.
- 1962 – Harry Hess studied the rocks on either side of the Atlantic Ocean and confirmed that the rocks got older on either side of the ridge. This supported sea floor spreading.
- More recent discoveries of areas of plate destruction.
- Earthquake epicenters outline the edges of tectonic plates.
- The youthfulness of ocean basins that are continuously being formed and destroyed.
- Global Positioning System (GPS) is now used to measure and record movements of plates and movements along faults.

Activity 12.1

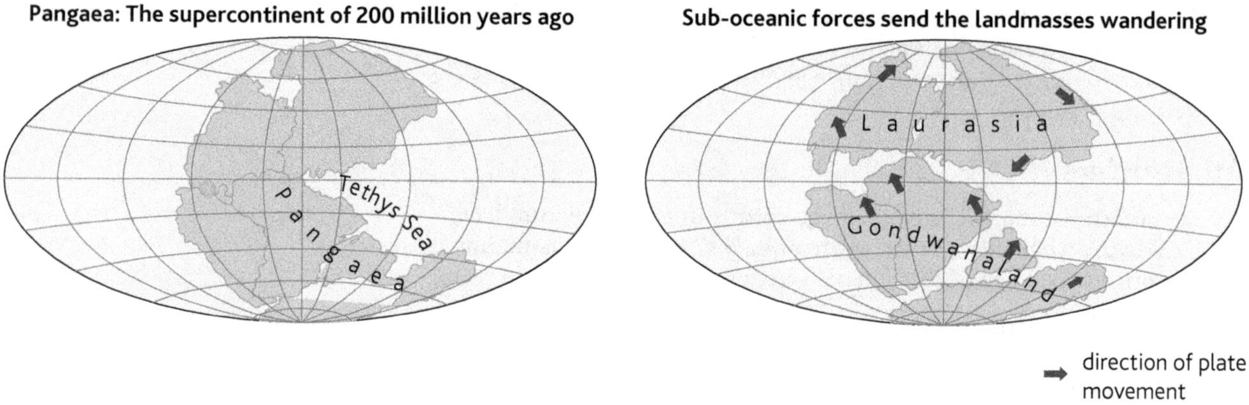

Figure 12.1

1 Name the theory to which Figure 12.1 refers.
2 Explain why scientists did not accept the theory referred to above.
3 Describe two pieces of evidence used to support the idea of the sea floor spreading.

Feedback

1 Theory of Continental Drift.
2 The theory was not accepted because scientists could not conceptualize of any mechanism or energy source which could cause and power continental movement.
3 One piece of evidence which supports sea floor spreading is the dating of rocks on the side of the mid-oceanic ridges. Research on either side of the mid-Atlantic Ridge showed parallel ages on either side.

   A second piece of evidence is the paleomagnetic record in rocks on either side of the ridge. The reversal of the magnetic poles in rocks on one side of the ridge is matched by the rocks on the other side of the ridge.

Formation, Distribution and Movements of Plates

The lithosphere consists of irregular segments of varying size and thickness called plates. The plates have definite boundaries or borders and these are found both on continents and oceans. The lithospheric plates are of two types – oceanic plates are thinner than most continental plates. They are also denser. Most of the major plates consist of both types of plates although some are named after the continents that lie within them. The large Pacific plate is almost entirely oceanic.

There are seven very large plates: Pacific, North American, South American, African, Eurasian, Indo-Australian and Antarctic; and very many smaller ones, such as the Caribbean, Nazca and Cocos.

The plates meet at different types of boundaries or margins and while three of these are well defined, there is a fourth area where movement is not clearly understood.

Activity 12.2

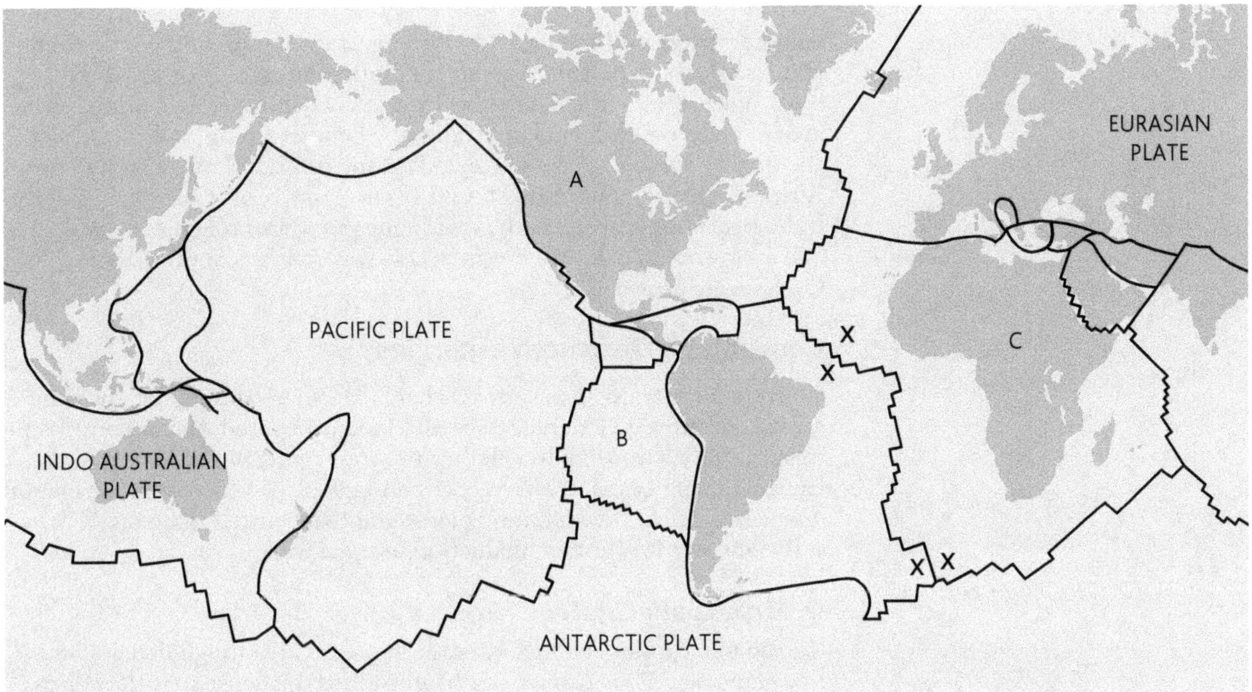

Figure 12.2

1 Plate A is the

- a North American
- b United State of America
- c Atlantic
- d African.

2 Plate B is the

- a South American
- b Andean
- c Cocos
- d Nazca.

3 The plates at places X-X are moving

- a apart
- b together
- c transform
- d up and down.

Feedback

1 a 2 d 3 a

Processes at Plate Margins

Divergent/Constructive margins

Plates are thought to be formed from upwelling of magma from the mantle onto the earth's surface forming new plate material. Some of these areas are marked by mid-oceanic ridges. The best studied of these ridges in the Mid-Atlantic Ridge stretching from the Arctic Ocean to beyond southern Africa. Basically it is a submerged ridge with a rift valley at its centre. Here, new crust is formed and plates move away from each other. Movement has been measured as averaging 2.5cm/year. At its northern end, volcanic Iceland lies over the ridge, splitting apart as the North American and Eurasian plates move apart.

Another divergent boundary has developed on the African continental plate marked by the East African Rift Valley. Stretching of the earth's crust under forces of tension causes it to crack producing a dramatic area of steep escarpments and valleys. The western rift valley contains the world's second deepest lake, Lake Tanganyika. Mount Kilimanjaro, Africa's highest mountain is found in the eastern rift valley. Ultimately if the spreading continues the continent would be broken into two parts and a new ocean created. It was in this way that the South Atlantic Ocean was formed.

Convergent/Destructive margins

Convergent margins mark areas where plates are coming together and crust is destroyed. The processes and landforms produced at convergent margins vary according to whether the converging plates are carrying continents or oceans. Convergence can take place between a continental and oceanic plate, two oceanic plates and two continental plates. The collisions result is either subduction or upheaval.

Oceanic-continental convergence

Continents cannot subduct because the rocks are of high density and low buoyancy. They cannot be absorbed into the dense mantle. When continental and oceanic plates converge, oceanic plates subduct leaving the continental plate floating on the asthenosphere. The extensive fold mountains of North and South America, the Rockies and Andes are formed from the meeting of the American plate and the Pacific and Nazca Plates respectively. In the case of the Andes, subduction and volcanic activity elevated the surface into two chains – what is now the Western Cordillera and further east, the Eastern Cordillera. The area between the two was filled with material eroded from the mountains to form the altiplano. Some of the most destructive earthquakes are associated with these regions. Volcanic eruptions are also common.

Oceanic-oceanic convergence

When oceanic plates meet, one subducts and is assimilated into the mantle. The older, heavier plate which is denser, will plunge beneath the younger. Older more rigid plates will also subduct at a sharper angle than younger. When the downgoing slab reaches a depth of about 100km there is partial melting which produces magma. This magma rises to the surface producing a chain of volcanic islands or an island arc. Trenches mark the position at which the subducting plate begins to descend beneath the lithospheric slab. Trenches are the deepest part of the ocean floor. The Mariana trench marks the area where the Pacific plate plunges beneath the Philippine plate. The Challenger Deep at the southern end of this trench is 11,000m deep. In the Caribbean, the North American plate dives below the Caribbean plate forming a subduction zone.

Trenches generally run parallel to volcanic island arcs. The Mariana islands, an archipelago consisting of 15 volcanic islands, lie to the west of the Mariana trench. In areas where rivers supply large volumes of sediment to the sea, trenches may not be apparent since they may be filled with sediments. The presence of the Orinoco River explains the absence of a trench along the island arc of the Lesser Antilles.

As the plate subducts, the overriding plate scrapes sediments as well as projecting portions of ocean floor off the upper crust of the lower plate. This creates a zone of deformed rocks that attaches itself to the overriding plate. This zone is known as an accretionary prism or wedge. Barbados to the east of the Lesser Antilles volcanic chain (Figure 12.3) is an emergent part of the accretionary prism.

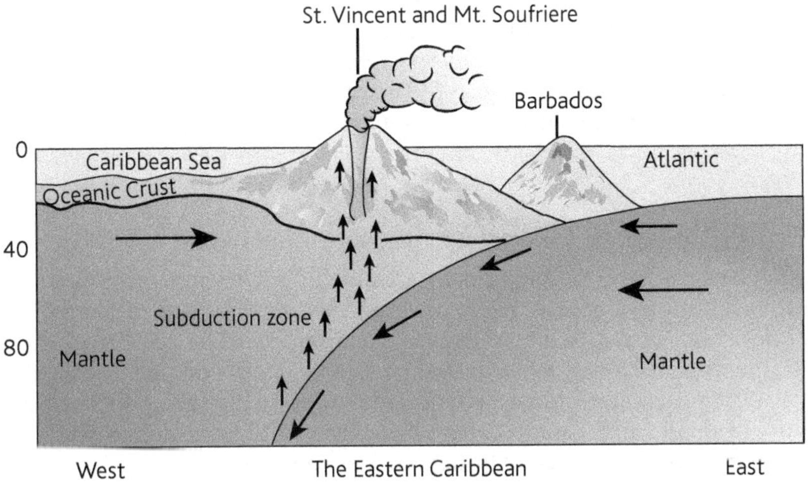

Source: Adapted from N Sealey, Caribbean World, Cambridge 1992

Figure 12.3 *Cross section through Eastern Caribbean*

Activity 12.3

1. Using evidence from Figure 12.3:
   a. Explain the formation of the island of St Vincent.
   b. Suggest how the island of Barbados was formed.
2. Describe the formation of three topographic features resulting from sea floor spreading.

Feedback

1. a. The diagram shows material (magma) rising from the area of subduction. This is material produced by the melting slab as the Atlantic Plate subducts beneath the Caribbean. The rising material produces a volcanic island such as St. Vincent.
   b. The diagram shows Barbados lying near the boundary of the two plates on the surface. The island is formed by the sediments – the accretionary wedge – scraped off the lower plates as it subducts.
2. Three topographic features formed from ocean spreading are mid-oceanic ridges; oceans and volcanoes. Mid oceanic ridges are submerged double range of mountains which mark the actual boundary of the plate material diverging, for example, Mid-Atlantic Ridge. Ocean basins are formed as the sea floor spreading breaks the continent apart and the resulting basin is filled with sea water, for example, the Atlantic Ocean. Volcanoes occur in areas of sea floor spreading as magma flows out on the surface to form islands such as Iceland.

Continent-continent convergence

When two continental plates converge there is no subduction of the light, buoyant material. Instead, the rocks crumple and buckle. The plates push under, into and over each other creating crustal thickening, folding, faulting, and overriding and some of the most complex structures in the world. The collision of the Indian and Eurasian plates has thrown up the Himalayas and the Tibetan Plateau. The Himalayas now rise to 8,854m. The Indian plate is moving northwards at a rate of about 5cm annually producing earthquakes that affect several countries in the region such as India, Pakistan and China. The collision of the African and European plates formed the folded Alps of southern Europe.

Transform/Neutral margins

In many areas of the earth, plates move laterally and slide past each other at transform faults. At these margins the crust is neither destroyed nor created. The fault surfaces are rough and pressure builds up as they lock together. The releasing of this pressure takes the form of earthquakes. Many of these transform faults are associated with ocean ridges, but in North America the San Andreas Fault lies on the surface in one of the most densely populated areas of California, USA. The fault which produced the earthquake in Haiti in January 2010 is also a transform fault.

Earthquakes and Faulting

Earthquakes can be created by undersea landslides, volcanic eruptions and the explosion of bombs. Most are explained on the basis of plate tectonics. They are caused by movements along a fault or fracture in the earth. Plates move away, towards and past each other along fractures or faults. Where the plates move apart there is tension rupturing rocks and producing small earthquakes. Where the plates slide pass each other, the rough edges lock, pressure builds up and energy is released when the stresses are overcome. Subduction and continental collision release large amounts of concentrated energy which results in some of the world's largest earthquakes. The ruptures produce shock waves which are felt as earthquakes.

The actual point on the fault where the rupture occurs is the focus and the point directly above it on the surface is called the epicentre. The epicentre experiences the strongest waves with decreasing intensity in concentric circles away from it.

When the epicentre of an earthquake is undersea, the energy passes through the water creating fast-moving waves travelling over great distances called tsunamis. In Thailand 2004, there was displacement along thousands of kilometres of an underwater plate margin, the great release of energy created tsunamis reaching right across the Indian Ocean to Africa hours later. Tsunamis are extremely hazardous to human life and property in densely settled coastal areas.

Earthquakes consist of waves of different speeds and lengths created by energy released at points in or on the earth's surface. Seismic waves are recorded on very sensitive instruments called seismographs.

The energy released in seismic waves may either pass through the entire body of the earth or along the surface only. These are body and surface waves.

Body waves are the faster and are subdivided into:

- Primary (P) waves, the first to arrive. They have a 'push/pull' motion (compression) and move through both solid and liquid. Their speed depends on the density of the material through which they travel.
- The slower, rolling secondary (S) waves pass through rock only. They sheer the material through which they pass. They are transverse waves because they vibrate the ground in a direction that is transverse or at right angles to the direction of movement.

The analysis of the paths taken by P and S waves, their travel time, the disappearance of the S wave at the core-mantle boundary, gave clues to the structure of the interior of the earth.

Surface waves travel near the surface of the earth and their amplitude decreases with depth. They are:

- *Love* waves are also transverse waves causing movement from side to side on a horizontal plane. They do not move through air or water.
- *Rayleigh* waves have an elliptical, counter clockwise motion and are very similar to water waves. They are the slowest of all waves.

These surface waves are responsible for the most damage. Each creates different stresses and hazards as they pass through the built environment.

The main shock of an earthquake may be preceded and followed by smaller shocks. There are foreshocks and aftershocks, and aftershocks may continue for years.

Earthquake magnitude (Richter) and intensity (Mercalli)

There are two different scales for classifying earthquakes: the Richter scale and the Mercalli scale.

- The Richter scale of magnitude measures the energy released by an earthquake by its amplitude. It is a logarithmic scale from 0–9 where a magnitude of 5 is ten times greater than one of 4. This does not mean that the shaking of a magnitude 5 earthquake would be 10 times greater than a magnitude 4. The energy would be dissipated over a wider area and over a longer time and therefore may affect more people. The Richter scale is an open-ended scale.
- The Mercalli scale from I – XII, describes the effect of the earthquake on the human environment. For example IX on the Mercalli scale is 'Damage considerable ... partial collapse of buildings' while XI is 'Few if any buildings remain standing. Bridges destroyed ...'. The Mercalli scale is considered less precise than the Richter scale, as it depends on factors such as perception of the observer and age/structure of the buildings.

There is a relationship between the two scales in that the greater the magnitude and the energy released, the greater is the likelihood of destruction.

Earthquakes occur widely over the earth's surface. Many are associated with plate margins, but they also occur in the centre of plates. In the Caribbean, most islands experience earthquakes: to the north the American plate moving westwards along a transform boundary creates earthquakes in Jamaica and the Greater Antilles; while to the south, the massive El Pilar fault between the Trinidad and South American plate results in frequent earthquakes. Earthquakes are rare in Barbados, lying east of the tectonic activity, but it did experience one in 2008. Guyana does not have frequent earthquakes as it lies on the very old, stable continental shield of South America.

In January 2009, Poas, Costa Rica experienced a magnitude 6.2 earthquake. This was not a very severe event but it had a serious social and economic impact on this vulnerable population. This impact will be examined in Module 13 – Floods, Earthquakes and Volcanoes as Hazards.

Faulting

As plates move the rocks fracture or crack. As stress builds up in the rocks on both sides of the fracture movement takes place and a crack is transformed into a fault.

Geologists describe a fault plane with two measurements – the strike and the dip. Strike refers to the direction of the fault on the earth's surface. The dip measures the direction and angle of the fault plane. The direction of the dip is perpendicular to the strike. Strike is seen in the horizontal and dip in the vertical view (Figure 12.4).

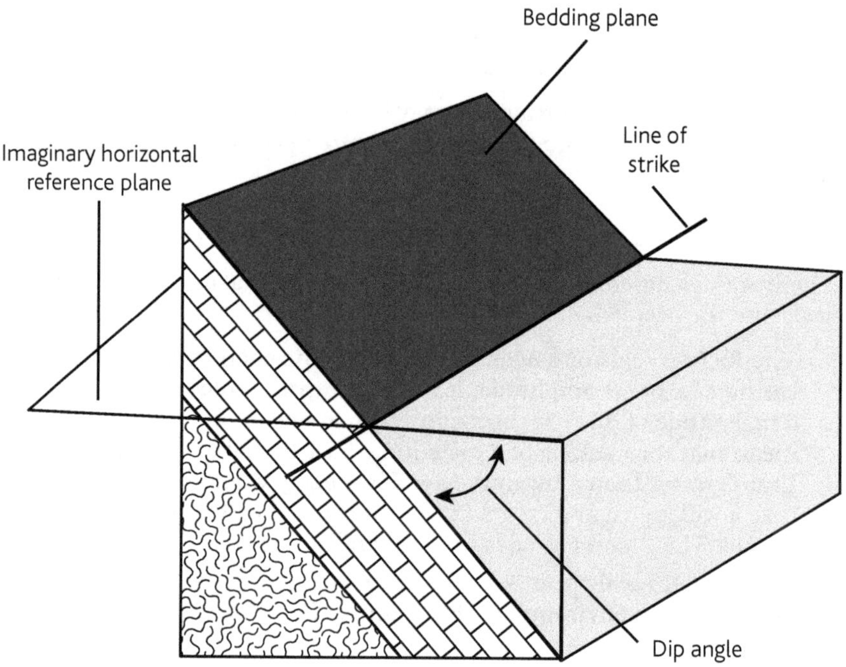

Strike and Dip symbols. Notice that they are oriented in the same directions as in the diagram above.

Source: http://web.arc.losrios.edu

Figure 12.4 *Strike and dip of rock layers*

The two sides of a fault are the *hanging wall* and the *footwall*. When the rocks on both sides of the fault change their vertical position a dip slip fault is formed. When the rocks move horizontally, a strike-slip fault is formed. There are no hanging and foot walls in strike slip faults such as the San Andreas fault.

The types of faults reflect the forces that are acting on the fault.

- Normal faults are created by the forces of tension and are typical of faults at spreading centres. The hanging wall moves down in relation to the footwall.
- Reverse faulting occurs where the forces of compression push the hanging wall upwards relative to the footwall. There is reverse faulting at convergence zones, where subduction or collision occurs.
- Strike-slip faults occur where the movement is parallel to the strike or horizontal. They are classified according to the relative direction of movement of the fault blocks. They may be right lateral or left lateral. If a person straddles the fault and the block on the right moves towards him/her then the fault is a right lateral or dextral fault. If the block on the left moves towards him/her it is a left lateral or sinistral fault.

- The most famous strike slip fault is the San Andreas Fault, California, USA. This is a right lateral systems fault that stretches over 1,200km. The Pacific plate is moving northwestwards, taking Los Angeles with it while the North American plate moves southeastwards relative to it.

Volcanoes

Vulcanicity refers to all the landforms derived from magma cooling within or on the earth's crust. Volcanic activity is associated with most plate margins although the specific form is related to the type of vent and type of lava.

Distribution of volcanic activity

Volcanic activity can be traced around the world to areas of tectonic activity such as plate margins. The 'Ring of Fire' of the Pacific (Figure 12.5) refers to the encircling or circum-Pacific volcanic activity at all margins of this large ocean. Volcanoes stretch from Aconcagua and Cotopaxi in South America; Popocatepeti and Mt St Helens in North America; through the Aleutians down the western Pacific in the Kuriles, Japan, the Philippines, Fiji and New Zealand. Other areas of active vulcanicity include the Eastern Caribbean, East Africa, Indonesia and the Mediterranean Europe. Volcanoes also form over 'hot spots' such as those forming the Hawaiian Islands (Figure 12.6).

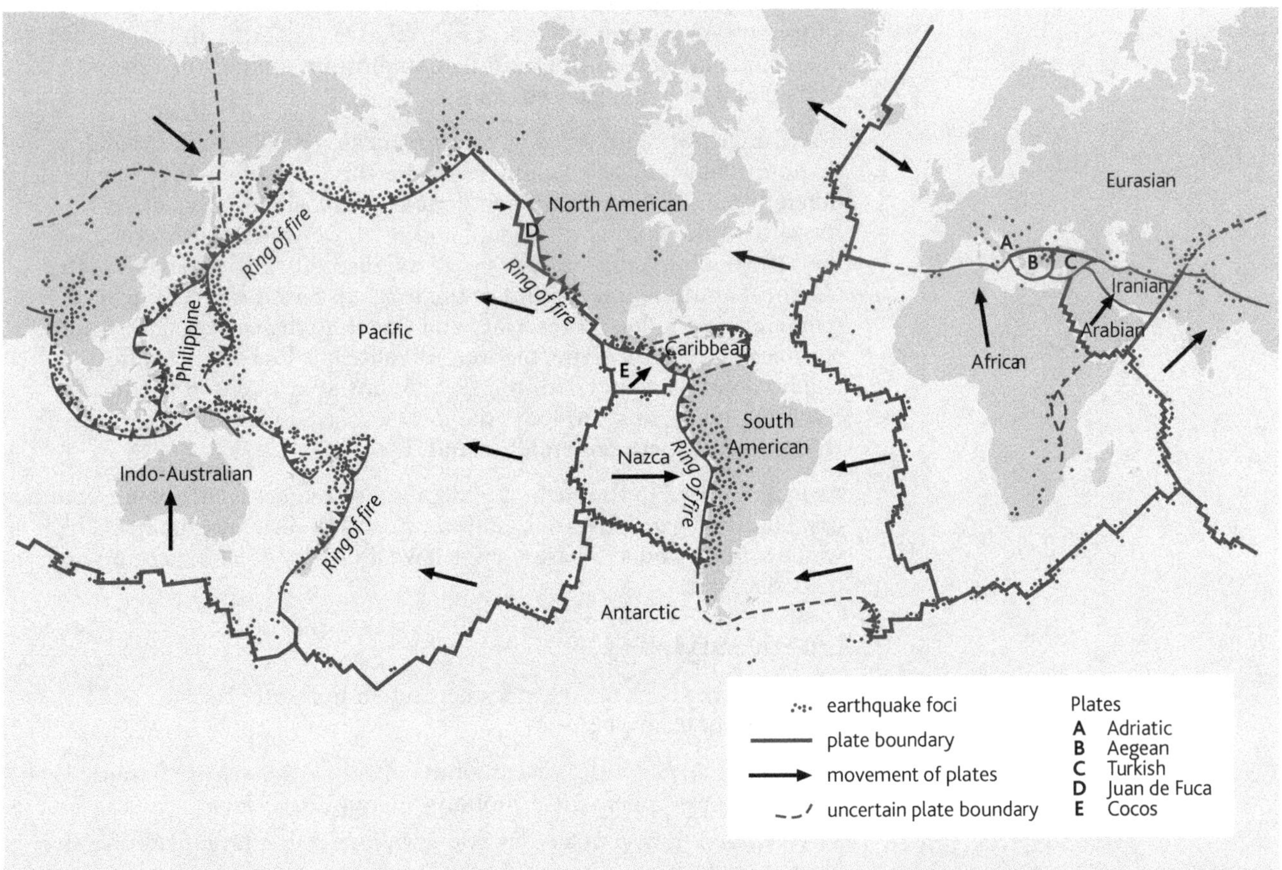

Figure 12.5 *Distribution of plate boundaries and volcanoes*

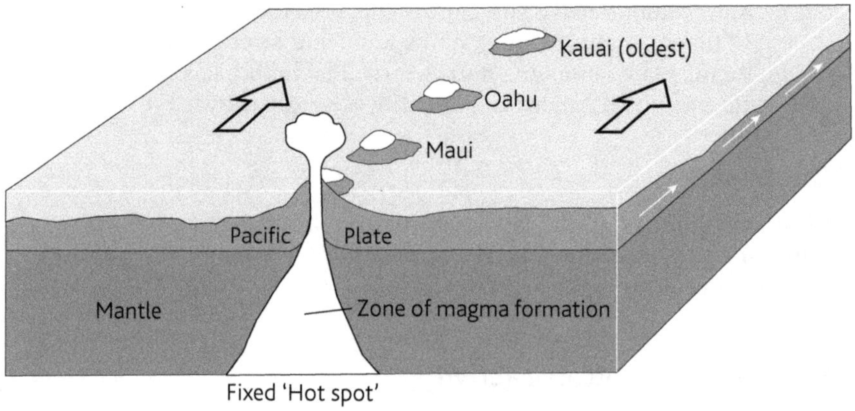

Figure 12.6 *Movement of the Pacific Plate over the fixed Hawaiian 'hot spot'*

Characteristics of volcanoes

Magma

Landforms (including volcanoes) produced by magma pouring onto the surface are called extrusive forms.

Some magma may not erupt onto the earth's surface but push into the underlying rocks. These produce intrusive features. Intrusive forms are not usually hazardous and are only identified when revealed on the surface by subsequent erosion. Their relative resistance influence rates of erosion and may form prominent landforms, such as the Pitons of St. Lucia, which are exposed pipes.

There are three main types of volcanic rocks – basalt, andesite and rhyolite – and they behave differently for three reasons – they have different temperatures, water content and viscosities. Viscosity is a measure of the fluidity of a liquid and the lower the viscosity the more fluid the liquid. Water has low viscosity. Viscosity also depends on the chemical and mineral content of the magma. Basalt has the highest temperature and lowest viscosity and therefore the easiest flow. Andesite occupies an intermediate position. Rhyolite has low temperatures and high viscosity and does not flow. About 80 per cent of the magma reaching the Earth's surface is the low viscosity basaltic and the rest shared equally between andesite and rhyolite.

As magma rises to the surface, and pressure decreases, the water becomes gas. Basaltic lava is low in dissolved water. Rhyolitic magma has a high water content and as it rises, gases have difficulty escaping because of high viscosity.

Eruptive styles

The eruptive styles of volcanoes depend on the water content and viscosity. For example:

- *Icelandic type:* with large amounts of very low-viscosity basaltic lava and low water content. Eruptions are non-explosive.
- *Hawaiian*: low-viscosity lava, low water, basaltic lava. Eruption style is non-explosive.

- *Strombolian*: relatively small amounts of moderately viscosity lava; moderate water content and usually peaceful; forming scoria or cinder cones.
- *Vulcanian*: high-viscosity lava; basalt to andesite lava; moderate to high water content and moderately violent eruptions.
- *Plinian*: very high-viscosity lava; basalt to rhyolite lava; high water content; very violent eruptions; strato volcanoes.

Volcanic landforms

- *Lava plateaus* are formed when the magma pours out of long fissures covering large areas with very fluid basaltic lavas. The eruption style is Icelandic, the most peaceful eruptions. This occurs at divergent plate margins where new plate materials are being formed, for example, in Iceland. The old Columbia plateau, USA and the north-west Deccan plateau are also lava plateau. Lava thicknesses of up to 1800m cover more than 400,000sq km of land in each case.
- *Shield volcanoes* are broad low profile features that form some of the largest volcanoes in the world. The lower slopes are gentle, middle slopes steeper, and summit flattened. Their shape is controlled by the fluid basaltic lava escaping through isolated vents or fissures in quiet, Hawaiian type eruptions. As the lava flows along the surface it takes two forms – the smooth, ropy pahoehoe which changes to the rough, spiny a'a, as the lava cools and loses gas.
- *Cinder cones* are formed when magma is ejected from the vent and cools. On its descent magma is deposited as ash and cinder. They form from the rapidly cooling rock fragments. The lava is of low viscosity and the eruption style is often Strombolian. Mount Etna in Italy is an example of a cinder cone.
- *Composite* or *strato volcanoes* are the typical high, steep cones formed from alternating layers of thick viscous lava and pyroclastic material. These volcanic cones are formed from basaltic to andesitic lavas which flow very slowly. The material ejected during the eruption may cover a very wide area. The composition of the magmas may vary from one eruption to the next and the eruption style may be Vulcanian and Plinian. The Plinian eruptions are powerful vertical eruptions which carry pyroclastic material high into the atmosphere. Smaller parasitic cones may develop on the sides and lava domes consisting of highly viscous material develop in the craters as in the Soufriere Hills. Mount St Helens in the USA, Kilimanjaro in Tanzania and Fuji in Japan are composite volcanoes.

(Tephra refers to all sizes of solidified magma particles which are blown out of the vent and fall to the surface under gravity. This includes the finest ash < 4cm; lapilli between 4–32cm; and larger bombs and blocks. Pumice, with its many holes, is a type of tephra.

Pyroclastic flows (nuee ardente) are very hot, fast moving clouds of gases and tephra racing down the side of a volcano).

These are all major hazards to human life and property as experienced in Martinique, Montserrat and other Caribbean islands. The impact of volcanoes on human life and property will be examined more closely in Module 13 – Floods, Earthquakes and Volcanoes as Hazards.

Activity 12.4

Use the diagram (Figure 12.7) to answer the following questions:

1 The area labelled A is a:
   a mountain root
   b magma chamber
   c parasitic cone
   d lava flow.
2 The arrow B points to:
   a old flow
   b magma
   c ash
   d lava.
3 The material labelled C is:
   a magma
   b cone
   c tephra
   d lava.
4 The type of volcano shown is a:
   a lava plateau
   b shield volcano
   c cinder cone
   d composite volcano

Feedback

1 b
2 c
3 d
4 d

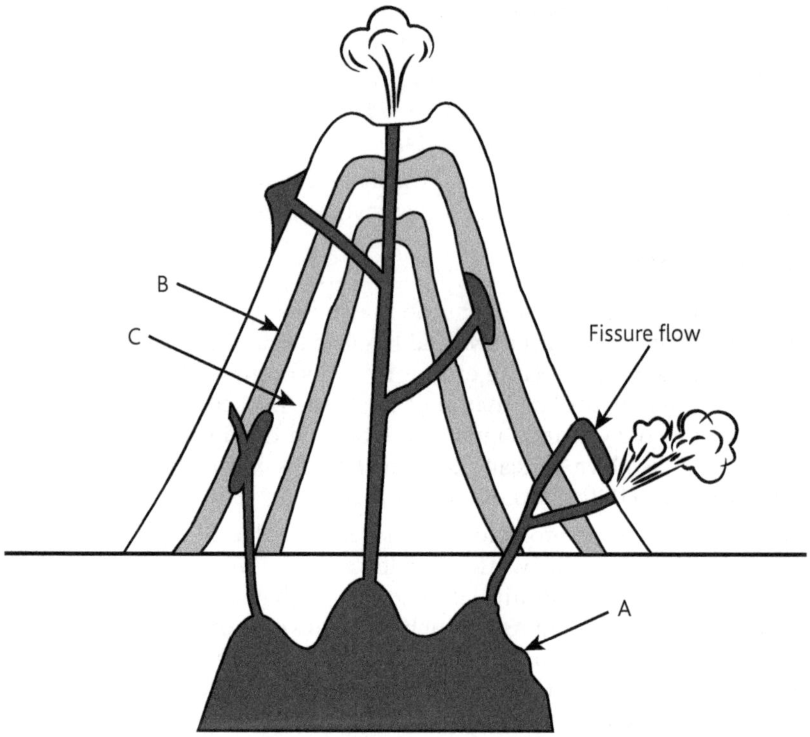

Figure 12.7 *Cross section through a type of volcanic cone*

Characteristics of Caribbean volcanic landscapes

The volcanic island arc of the Caribbean extends from the Virgin Islands in the north to those off the coast of Venezuela. There are about 17 active volcanoes in the chain including the submarine volcano, Kick-em-Jenny to the north of Grenada. Volcanoes such as the Soufriere Hills of Montserrat are very complex andesitic strato volcanoes with lava domes forming the summits. These active volcanoes create a great variety of landscapes: some islands, like Nevis, are formed from a single volcanic cone resulting in a steeped sided mountain of radial drainage; others like Dominica have a complex of nine active volcanoes producing overlapping steep slopes.

Most Caribbean volcanic islands are characterized by their steep slopes rising to heights of over 1000m above sea level. Many are topped by water filled craters.

Postive Impact of Volcanic and Earthquake Activity

Both volcanic eruptions and earthquakes are extreme natural events that are often hazardous to human life and property. However they both have positive impacts. They are both important to human survival.

Volcanic activity

Volcanic activity has positive impacts in the following ways:

- *Fertile soils*. The volcanic material once cooled and weathered forms rich soils. These soils support the agriculture of many densely settled areas such as in the Mediterranean and the Caribbean islands.
- *Land surface*. Erupting volcanoes are thought to be the source of the first land as the hot planet cooled. Eruptions continue in the sea

creating new land. All the volcanic islands are created in this way. The shape of Montserrat has changed when the last eruptions extended the land in the south. (But more than a half of the land area is now uninhabitable.)

- *Minerals*. Many useful materials are formed directly or indirectly from volcanic activity. Building materials such as granite and marble, precious minerals such as diamonds as well as silver and copper are formed from magma in and on the earth's crust.
- *Geothermal energy*. Some volcanic countries have tapped the natural heat energy of volcanoes as a source of electricity.
- *Tourism*. Volcanic peaks, whether active or dormant are natural tourist attractions in many parts of the world. From the majestic Mt Fuji in Japan to the ever flowing Hawaiian cones, tourist visit and wonder at this evidence of the earth's interior.
- *Science*. Volcanic activity allows scientists to study the materials of the interior earth's crust. This is very important since it is too hot and far from the surface for direct investigation.

Earthquakes

Earthquakes are more commonly known for their devastating impact on human life and property, but they also have a positive impact. Earthquakes allow scientists to better understand the nature of the earth's crust. Their passage through different layers of the earth's materials provides valuable data. Indirectly, earthquakes also result in improved building construction.

Value of Folded and Faulted Landscapes

- Sources of minerals, for example, coal beds in the Appalachians, as well as minerals intruded below the mountain as batholiths, for example, tin in the Andes.
- Energy – water falling over high relief has been used for hydroelectricity, for example in Norway and Canada.
- Tourism – high folded mountains have been used for both summer and winter tourist activities, such as skiing, hiking and mountain climbing. For example, Swiss Alps.
- Biodiversity – many high fold and block mountains have developed their own unique flora and fauna. As they are less accessible than lowland areas they remain as reserves for scientific enquiry.
- Country/Regional boundaries – high mountains often serve to separate one area/country from another.

Key Points

- Earthquakes, folding, faulting and volcanoes are associated with plate tectonics and are natural hazards.
- The earth's surface is made up of crustal plates of different sizes.
- There are three distinct margins: Convergent, Divergent and Transform.
- Continents and oceans lie on plates and move with them.
- Plate tectonics helps to explain the process causing earthquakes, folding, faulting and volcanic activity.

Conclusion

It is important to understand tectonic activity as natural events in order to be able to respond to and mitigate their impact as natural hazards. The processes associated with plate tectonics help to explain the formation and distribution of many of the earth's natural events which threaten human life and property. An understanding of plate tectonics can assist in the prediction and mitigation of tectonic hazards, such as earthquakes and volcanoes.

End Test

1 Compare divergent plate margins with convergent plate margins under the headings: Movement; Landforms; Volcanic activity.

2 Explain why some plates subduct and others do not.

3 Give two reasons why flooding may follow the eruption of volcanoes.

4 Describe normal and reverse faults in terms of movement of blocks and forces involved.

End Test Feedback

1 *Movement*: Divergent plates move away from each other while convergent plates come together. Magma comes up to fill the space between the diverging plates resulting in sea floor spreading. In convergence, oceanic plate material is subducted and destroyed.

Landforms: Divergent boundaries are marked by undersea mid-oceanic ridges but on the surface volcanic material may emerge as islands. Convergent margins are marked not only by deep sea trenches and volcanic island arcs but also complex folded and faulted features of compressed/broken continental materials.

Volcanic activity: At divergent boundaries, magma erupts through mainly fissures with little or no explosive activity; whereas at convergent margins magma escapes through vents which are easily blocked by acidic lavas leading to explosive eruptions often including gases.

2 Continental plates do not subduct because they are composed of light buoyant material which cannot be absorbed into the dense mantle.

3 Flooding may follow volcanic eruptions because of heavy rainfall in the area. Steam released in the air may add to the likelihood of rain. In other cases, magma may melt the ice of glaciers in the areas and create floods.

4 In normal faults the force is usually one of tensions causing the hanging wall to move downwards in relation to the footwall. In reverse faulting compressional forces push the hanging wall upwards. One side of the fault plane overhangs the other.

13 Floods, Earthquakes and Volcanoes as Hazards

General objective

On completion of this module you should appreciate the relationship between the natural and human environments.

Specific objectives

You should be able to:

describe the impact of floods, volcanoes and earthquakes as hazards

use photographs, hazard risk maps and topographical maps to analyze vulnerability to hazards.

Introduction

Human beings live all over the world even in hazardous areas, since the positive aspects may outweigh the potential hazards. Extreme events which occur in uninhabited areas of the world are not hazards. Each natural event can have several impacts on human life and property. Some events take lives and destroy property directly (primary effect); while others have indirect effects (secondary effect). Tertiary effects are long term effects which may be permanent.

Flooding has many negative impacts on human life and property. Flood waters can sweep people to their death and cause damage to property. This is the most common hazard experienced globally and throughout the Caribbean.

Earthquakes are the most sudden in onset, striking with little warning. The violent seismic waves immediately destroy buildings and kill persons by burying them under the rubble. They may trigger landslides and damage dams. Tsunamis may be created by undersea earthquakes and flood coastal areas.

Volcanoes are perhaps the most dramatic extreme natural event. Hot lava spewing out of the vent; ash and gases exploding into the air; hot clouds of gases speeding down slopes, are very spectacular occurrences. The impact can be devastating, obliterating large areas with people and buildings buried under tons of hot ash/lava. Some volcanic eruptions are not explosive but can still have a negative impact on the human environment.

This module will look at the impact of these three types of hazards on human life and property. Module 14 – Response to Hazards will examine our response to these events, in prediction, mitigation and preparedness strategies. Responses will be examined at individual, collective and government levels in that module.

Content

- Hazard impacts.
- Consequences of flooding.
- Factors influencing earthquake impact.
- Nature of volcanic hazards.
  [Case studies included]

Hazard Impacts

Hazard risk arises from the intersection of 'natural events' and 'vulnerable population' (Module 11 – Natural Events, Hazards and Disasters – Flooding). The magnitude, speed, extent and duration of the event, all influence its impact, but the characteristics of the human environment will also play a big part in the realization of the hazard, the disaster or the catastrophe.

Generally the impact of any hazard is dependent on the number of people and type of human activity in the affected area. Populations in LDCs are often unprepared and ill equipped to respond to hazard impacts. Many lives are lost and it is often difficult for communities to return to previous living standards. In MDCs, there are greater property losses than loss of life as higher levels of technology and resources allow more people to prepare, evacuate or recover from the effects of hazards. For example, flooding of the Mississippi river in the USA has a different level of impact from flooding in the Ganges delta where the population is larger and more vulnerable.

The magnitude of an event will directly influence its impact: the greater the size of the event, the greater the hazardous effects. Its frequency will also affect how it affects people. Generally people are better prepared if they have experienced an event. Planned land use zoning and evacuation may lessen the impact. Table 13.1 shows some other characteristics of the hazard events addressed in this module.

Table 13.1 *Hazard event characteristics*

	Flood	**Earthquake**	**Volcano**
Speed of onset	Moderate	Fast	Moderate
Duration	Days/weeks	Seconds	Days
Area	Widespread	Concentrated	Limited

Each natural hazard can have primary, secondary and tertiary effects depending on the characteristic of the population and the level of development.

- Primary effects are those caused directly by the event – for example, flood waters may sweep away houses, or earthquakes may cause buildings to collapse.
- Secondary effects are those in the aftermath of the event, for example, famine, disease or fires after the event is over.
- Tertiary effects are long term/permanent changes, for example, relocation of settlements.

Consequences of Flooding

Flooding impacts:

- People are washed away and drown.
- Animals, cars and buildings are washed away.
- Property and crops are destroyed by inundation.
- The flood waters often leave deep layers of mud on flooded surfaces destroying buildings and their contents.
- Flood waters can contaminate drinking water supplies and cause diseases.
- Destruction of underground utilities.

Case study of flooding in Guyana

In January 2005, Guyana experienced its 'worst natural disaster'. In that month, the country received 1268mm of rainfall when the average amount of rainfall in January was 178mm. In one night there was 171.02mm of rainfall. This was in addition to the saturating rainfall of late December 2004. The government declared a disaster as flood waters covered the low-lying coastal areas of West Demarara/Essequibo Islands, Demarara/Mahaica and Mahaica/West Berbice. This narrow coastal strip is very densely settled, containing the capital and supporting the main agricultural lands. Of a total population of 750,000, 70 per cent live in this coastal area. It was estimated that this event directly affected about 290,000 persons with flood waters rising to 1–2m in affected areas. Ninety two thousand persons had their homes flooded; houses and their contents were mud covered as the waters receded; 5,600 persons were evacuated to 43 shelters and 32,000 were without access to assistance. Water over-topped the large reservoirs at East Demerara flooding the surrounding villages. Secondary effects included food and water shortages. Thirty five persons died, 21 from an outbreak of leptospirosis.

The mean gradient of the major rivers is about 0.20 and in many areas river flow is sluggish. An extensive drainage network of canals, conservancies (dams) kokers and sluice gates (sea defenses) cover the agricultural area. But these had been poorly maintained and in some areas they were not functioning. The rising flood waters could not be released quickly enough because of inadequate pumping facilities and blocked drains. Guyana is one of the poorest countries in the region with a GDP per capita of about US$ 1,200 per year. Many years of deforestation in the upper reaches of the river basins and urbanization in some areas have added to the vulnerability of these areas. Guyana experienced the highest rainfall since record keeping began in 1888 and the deluge caused havoc among a vulnerable population.

Activity 13.1

Photograph of area affected by flooding

Study Figure 13.1, which is a photograph depicting some consequences of severe flooding in Gonaives, Haiti caused by the tropical storm Jeanne in 2004, and answer the questions below.

Figure 13.1 *Some consequences of severe flooding*

1 Identify three types of physical damage evident in Figure 13.1 above.

2 Outline the main environmental risk associated with severe flooding.

Feedback

1 The photograph shows physical damage to:

   a vehicles – some of which have been overturned

   b houses – one has been completely demolished

   c residual mud left by the floodwater has caused damage to the area around the houses. There will also be substantial damage inside the houses.

2 The main environmental risk with severe flooding is the spread of water-borne diseases. Flood waters are easily contaminated by the debris, dead animals and organic matter which float along in the water. After the water recedes all items are covered with mud and silt. The mud itself can carry bacteria and as it dries can release harmful dust into the air.

Primary and Secondary Effects of Earthquakes

The seismic waves generated by earthquakes are most powerful at the epicenter. Depending on the type of material they pass through and the type of relief, they may have many effects. Although earthquakes last for a few seconds they present great hazards particularly in the built environment of tall densely populated buildings.

Earthquake impact may include the following *primary* effects:

- people may be killed by collapsed buildings
- destruction of roads and bridges
- liquefaction of alluvial material
- landslides and avalanches
- ruptured underground gas/electricity/water pipes
- fissures open in ground
- trees toppled
- tsunamis.

Secondary effects include:

- homelessness
- fires
- floods
- famine
- diseases
- disruption of waste and sewage disposal systems
- lack of potable water.

These secondary effects can cause greater loss of life in remote or isolated areas. People die of exposure in cold mountainous winter weather, or die of starvation because aid cannot reach them promptly.

Tertiary effects may include permanent evacuation of the area and changes in relief, for example, displacement of land along faults.

Earthquake impact in Poas, Costa Rica, January 2009

The magnitude 6.2 earthquake with its epicenter at Poas, Costa Rica was not the most deadly but illustrates how the vulnerability of the population can influence the hazard impact as much as the magnitude of the event. The earthquake was responsible for 23 deaths, injury to 100 persons and the complete destruction of more than 400 houses. At least 71 communities had damage to their water and road supplies. In addition, Poas, which is a tourist destination and an important agricultural area, suffered economic damage. A hydroelectric plant, hotels, crops as well as schools and health centres were all affected by the earthquake.

Costa Rica is a LDC of 4.5m people, with a GDP per capita of less than US$5,000 per year. It lies on a very narrow volcanic mountainous section of the subduction zone between the Cocos and Caribbean plates. This is a very active tectonic zone.

Haiti

One of the largest earthquakes in the western hemisphere occurred in Haiti, the poorest country in the western hemisphere, on January 12, 2010. The magnitude 7 earthquake occurred along the Enriquillo Plantain Garden Fault, a strike slip fault which extends through the Dominican Republic, Haiti and Jamaica. In this region, the Caribbean plate is sliding to the east while a smaller Gonvave platelet, one of the large number of platelets between the Caribbean and North American plates, is moving westwards. It had its epicenter in the town of Léog ne and by January 24, about 52 aftershocks, some with a magnitude of 6 were recorded.

The *primary* effects were devastating. Buildings collapsed killing an estimated 200,000 persons. It was estimated that in the epicenter about 90 per cent of the buildings collapsed. Homes, hospitals, schools and the Presidential palace were damaged or destroyed. Many of the buildings were poorly constructed and not designed to withstand earthquakes.

Secondary effects included widespread homelessness. Infrastructure such as roads were damaged and many could not get to hospitals or secure treatment in hospitals and subsequently died. Slow distribution of food and other resources resulted in foraging and sporadic outbreaks of violence.

Activity 13.2

1 Earthquakes negatively impact on physical, man-made and human environments. Explain and give examples of each type of impact.

2 Suggest three reasons why 'the critical period of need is the first 24 hrs' after an earthquake.

Feedback

1 The impact of earthquakes on the physical environment includes changes to the landscape, as well as the occurrence of landslides and avalanches. For example, in Poas, Costa Rica, many landslides were triggered by the earthquake of 2009.

The man-made environment is sometimes also called the built environment that consists of buildings, roads and bridges which humans construct. These are often severely damaged by earthquakes. Buildings may be completely demolished while bridges and roads are impassably dislocated and broken. This also occurred in Costa Rica 2009 earthquake in the steep terrain.

The human environment and all aspects of social and cultural life are affected by earthquakes. As people mourn loss of life and possessions there is great social disruption and loss of community. In Poas, person's livelihood in agriculture was impacted as well as lack of water and food.

2 The critical period of need after an earthquake is within 24 hours since this is the small window of opportunity to rescue persons and also control the potential for secondary hazards. In the first 24 hrs it is possible to rescue persons buried under buildings. After 24 hrs without water, survival is unlikely.

Homeless persons exposed to the elements will begin to suffer the effects of that exposure. Already traumatized and in shock those persons will be even more vulnerable to prolonged psychological effects. Therefore, it is very important to get aid in the first 24hr to rescue possible survivors and stablise the situation of affected communities.

Volcanic Hazards

The eruption of a volcano can be one of the most dramatic hazard events although not the most common or necessarily the most destructive. Generally the hazardous conditions produced by a particular volcano are dependent on the type of volcano and the characteristics of the material ejected. These were examined in Module 12 – Plate Tectonics. The most hazardous eruptions tend to be those of viscous lavas accompanied by gases as in the Caribbean.

Volcanic hazards include lava flows, ash and volcanic tephra, pyroclastic flows, gases and acid rain and lahars. The impact of these hazards include:

- death by incineration, asphyxiation or burning
- destruction of buildings
- damage to vegetation
- destruction of all forms of transportation
- damage from mudflows.

An erupting volcano can literally bury the entire area under tons of hot lava and ash completely obliterating the previous physical and human environment. Most active volcanoes are monitored by seismographs and evacuation ordered but even previously extinct/dormant volcanoes can become suddenly active.

Volcanic activity in Montserrat

The impact of volcanic eruptions such as Montserrat is very well documented. The former lush green 'Emerald Isle' of about 11,000 persons mainly living around the capital Plymouth was transformed by the 1997 eruption into a small settlement of 5,000 persons restructuring their lives in the north at the new capital Brades. Nine persons lost their lives and the economy based on agriculture and tourism was destroyed. The entire southern half of the island, including the air and sea ports were covered in ash and lava and had to be abandoned.

The potential for volcanic hazards exists in all the volcanic islands of the Eastern Caribbean. For example, Dominica has nine active volcanoes which have not erupted in the last 500 years. The submarine 'Kick 'em Jenny' off Grenada could cause tsunamis reaching non-volcanic islands like Barbados.

Key Points

- Floods, earthquakes and volcanoes are significant natural hazards to human life and property.
- All hazards can have effects beyond the event itself (primary effect).
- Secondary and tertiary effects can be as devastating as the event itself.
- A single hazard event can generate many hazardous conditions.
- Most areas of the world have at least one hazard threat, but living in these hazardous areas can have its advantages. For example, the fertile soils found in volcanic and flood plain areas and fishing grounds in coastal areas that may be hazardous because of tectonic activity.

Activity 13.3

1 Gas emissions, lahars and pyroclastic flows are events associated with volcanic eruptions. **Research**: Describe the effects of these events on the environment.

Conclusion

As human beings cover the world with settlements and economic activity it is inevitable that there would be hazard risks. Every year many thousands of people and large parts of the environment come under the impact of floods, earthquakes and volcanoes.

End Test

1 Briefly describe two ways in which human activities can increase the effects of flooding.

2 Study Table 13.2 which shows the number of deaths and injuries resulting from earthquakes and volcanoes during the period 1972 to 1996.

Table 13.2 *Death and injuries from earthquakes and volcanoes*

Year	Earthquakes		Volcanoes	
1972–1996	Deaths	Injuries	Deaths	Injuries
	18 715	27 012	1017	285

Suggest two reasons for the differences in the number of deaths and injuries resulting from earthquakes and volcanoes.

End Test Feedback

1 Human activity can increase the effects of flooding by deforestation and urbanization. Deforestation in the river basin reduces infiltration and increases runoff so that more water flows on the surface. In addition the lack of vegetation also increases the rate at which the water moves downstream. Urbanization covers the soil with impermeable concrete surfaces which allow maximum runoff increasing the surface water. In addition the building of storm drains can increase the volume and velocity of runoff water. This increases the likelihood of flooding.

2 Two reasons for the differences in deaths and injuries shown in the table above are i) Earthquakes are much less predictable than volcanoes. They have a very rapid onset and can occur in very densely populated areas. Active volcanoes are better monitored and have a longer onset period in which people can be evacuated to safety. ii) Earthquakes can affect larger areas than volcanoes and so affect more people. Seismic waves spread out from the epicenter and affect people further away. Volcanoes tend to have a more localized effect.

14 Response to Hazards

General objective

On completion of this module you should appreciate the relationship between the natural and human environments.

Specific objective

You should be able to assess the responses to hazards.

Introduction

Responses to natural hazards, such as floods, earthquakes and volcanic activity, vary from an acceptance of the effects of the natural forces to complex technological prediction systems and elaborate preparations in the human environment. Generally four aspects of hazard management are recognized: Prevention and mitigation involves action to reduce the potential hazard impact; Preparedness is equipping people to cope with the hazard before it occurs; Response is an effort to reduce the impact of a disaster that has occurred; and Recovery contains efforts to restore lives to a normal state (Figure 14.1). Mitigation and preparedness are usually undertaken before the event; while response and recovery are done after the event.

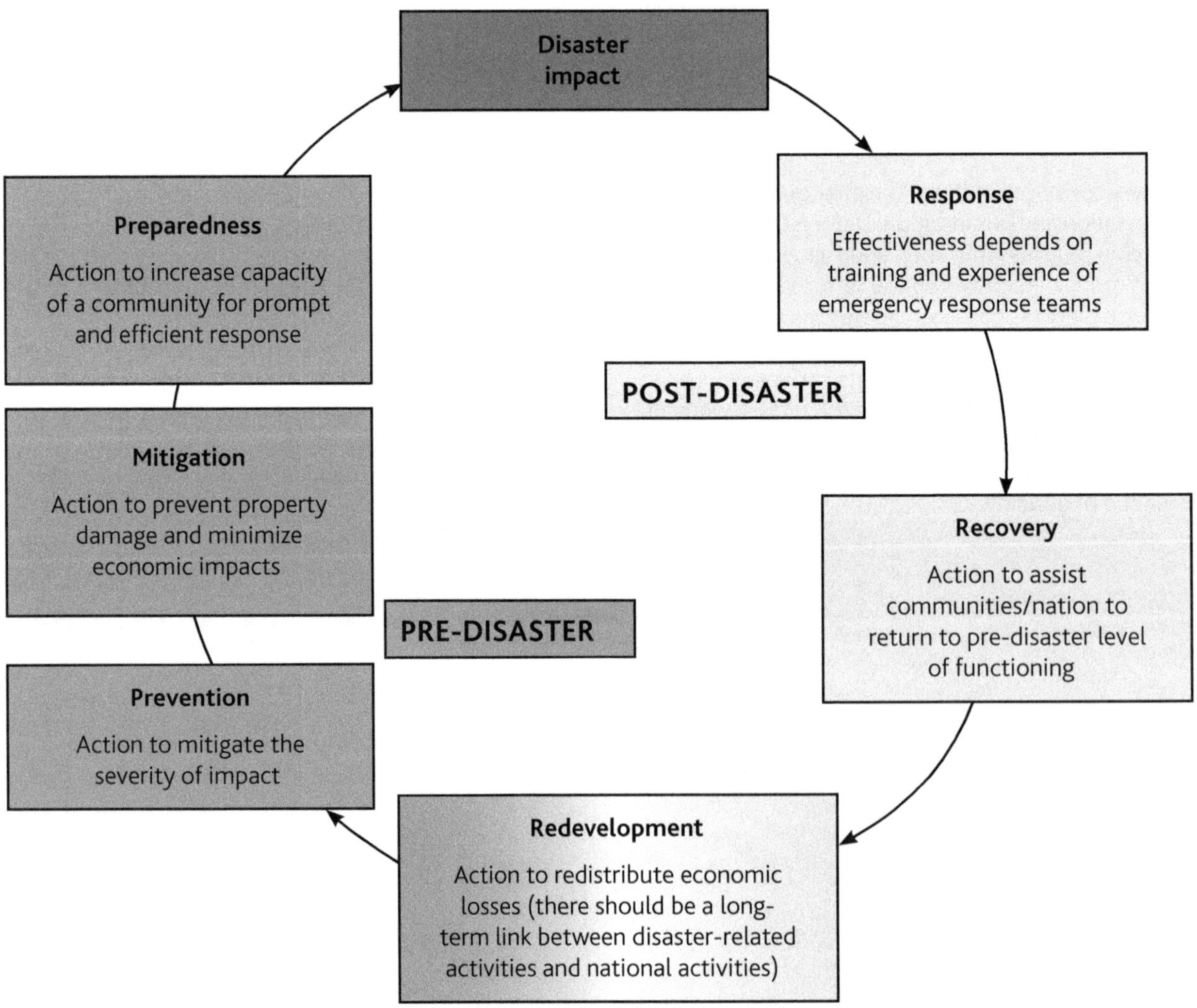

Figure 14.1 *Disaster management cycle*

Each hazard has its specific threats and characteristics as discussed in Module 13 – Floods, Earthquakes and Volcanoes as Hazards. Prediction of an extreme event is better in some hazards, such as flooding, than others, such as earthquakes. Human perception of risk varies with the perception, experience and awareness of the population. The response is also affected by the affluence of the community: LDCs struggling to provide basic needs for their populations are most vulnerable as a large proportion of the population lives in flimsy structures on marginal land. MDCs are better able to manage their hazard risk with mandatory evacuations; land use zoning and technological monitoring.

Effective preparedness can reduce the effects of disasters even for those who live in the most hazard-prone areas, lacking resources to meet the challenges of the recovery phase.

Content

- Prediction of flooding, earthquakes, volcanic activity.
- Individual and collective responses to floods, earthquakes and volcanic hazards.
- Government responses to flooding, earthquakes and volcanoes.

Predicting Floods, Earthquakes and Volcanoes

Floods

Flooding is perhaps the oldest and most predictable of hazards. Improved weather forecasting and river management methods make this the most predictable of hazards. However this does not lessen its impact because of the overwhelming attraction of low-lying coastal and valley areas to human populations. Certain data are needed to predict/forecast river flooding:

- The volume of rainfall and the location of the event.
- The intensity and duration of the event.
- The rate of change of water levels in rivers.
- The characteristics of the land area, for example, soil moisture conditions.

Recurrence interval data give information on the frequency with which floods of a particular magnitude are exceeded. It is useful when the aim is to provide structures, such as bridges, that would withstand the most extreme events. However, recurrence interval data are probabilities and are not important components of a flood warning system.

Prediction methods include:

- *Stream gauging* – precipitation and river flow are measured and monitored by gauges. Real-time (as they occur) precipitation data are used to forecast the height of water above a reference elevation (the stage). Data may be fed into a computer. The information is used to forecast floods.
- *Satellite imaging* – This involves the use of Earth Observation data. Radar images provide information on soil moisture conditions and flood extent.
- *Computer models* – Where the records of river discharge are available, mathematical models are developed on how rivers and streams would react to rainfall and snow melt. These are developed for selected points along rivers or in urban areas with a history of flooding. When

heavy rains are forecasted, the amounts are entered and the model estimates the resulting stage and discharge.

Earthquakes

The very sudden onset of earthquakes makes them difficult to predict. Most methods of earthquake prediction rely on inference from changes observed in the environment.

Prediction methods being tried:

- earthquake probability data
- strain accumulation
- animal observation
- rock deformation
- satellite data on ground tilt
- laser geodimeters
- radon gas accumulation
- magnetometers.

Volcanoes

Volcanic eruptions can be predicted by the earthquakes produced by the rising magma. However, volcanoes which were thought to be extinct can suddenly and unpredictably spring to life explosively.

Volcanic activity is monitored by:

- seismometers
- satellites
- tiltmeter
- ultra sound monitors
- gravimeters
- chemical sensors.

Individual and Collective Responses to Flooding, Earthquakes and Volcanoes

Government responses

Since the UN International Decade for Natural Disaster Reduction of the 1990s, most governments have undertaken to implement long and short term measures to cope with hazards. Many countries have national and regional organizations which are supposed to prepare and implement programmes to cope with hazardous situations. In the Caribbean, CDEMA (Caribbean Disaster Emergency Management Agency) coordinates regional disaster management, including response and recovery efforts. Individual countries have national organization such as the Central Emergency Relief Organization (CERO) of Barbados; the Office of Disaster Preparedness and Emergency Management (ODPEM) of Jamaica.

In much larger countries such as the USA, there are not only national organizations such as FEMA (Federal Emergency Management Agency), but also state institutions dedicated to the particular hazards for example, the Hawaii Tsunami early warning system.

One of the problems facing many governments is the fact that in many areas of the country, they may have to respond to the possibility of not just one hazardous situation but multiple. In many countries, informal settlements develop in areas that are avoided by the middle classes precisely because of the environmental problems posed – gully banks, river banks, abandoned water courses and steep slopes. In such situations they risk the effects of earthquakes, hurricanes, floods and landslides. In addition, many Caribbean countries face risks from earthquakes, volcanic eruptions and floods. Government strategies, therefore must be comprehensive.

Risk assessment

One of the first steps that ought to be taken is a comprehensive risk assessment. This involves:

- An identification of the nature, location and probability of the hazard/s.
- Measures of vulnerability – an assessment of who or what is exposed to the risks.
- As assessment of the resources that are available to reduce the risks, that is, institutional capacity.
- A risk analysis to determine the levels of risk.
- A risk evaluation designed to decide on the interventions and to establish priorities.

Today, this is facilitated by GIS technology which could provide a data base of disaster related information. This assessment must precede mitigation and preparedness measures.

Mitigation and preparedness

Preparedness involves those measures undertaken to ensure a readiness to forecast and respond to a disaster such as testing of early warning systems and training. Mitigations are measures aimed at reducing the impact of a disaster. Preparedness reinforces mitigation and sometimes no distinction is made between the two.

The Caribbean Disaster Mitigation Project (CDMP) has developed a methodology employing these assessment steps for use in the region and has been giving assistance to countries to improve their capabilities. The region has benefited from initiatives under each of the steps outlined:

- Hazard assessment and risk mapping – pilot projects in storm hazard modelling, assessments of landslides and earthquakes in Jamaica.
- Vulnerability – vulnerability audits for shelters in the Eastern Caribbean, community level vulnerability in Haiti and the Dominican Republic.
- Risk analysis – plans for the US Virgin Islands, Barbados, St Lucia.
- Intervention/implementation – development of building codes and guidelines.

Mitigation activities may involve:

- Structural measures, that is, construction to reduce or avoid impacts. It may apply to the design of new or retrofitting of the old – housing design; roof design, material and reinforcement. It also applies to the construction of levees, floodwalls and channelization.
- Non structural measures including government policies such as land use regulation, insurance, tax exemptions for risk avoidance, plans for evacuation, systems for monitoring, warning, education, training and acceptance of loss.

Legislation is necessary to promote both structural and non structural measures, for example, covering environmental protection and protection of watersheds; performance standards for buildings and land use zoning.

Response

In this stage, governments are usually assisted by NGOs such as the Red Cross and depending on the scale of the disaster; they may also need international assistance. The response stage covers events that take place immediately before (depending on the disaster) and after the event. Plans made in the preparedness and mitigation stage must now be put into action. Immediately after the disaster, first responders are sent out to the area to assess the damage. They are responsible for making preparations for receiving equipment and the volunteers who would rescue, provide food, organize shelters, evacuate and provide medical attention, security and counselling to those in need.

Recovery

The recovery stage is the longest. In the immediate aftermath, emergency relief must be provided for those who survive and who cannot provide for themselves. The provision of shelter is of extreme importance. Survivors are often in need of clothes and food. Long term recovery may extend over years. Five years after the earthquake in Pakistan, survivors were still living in tents. Permanent shelter must be provided for those in need; infrastructure must be restored – roads, water and sanitation services; decisions made as to whether whole communities should be relocated. Many are also in need of emotional support.

Community responses

Quick responses can make a big difference to the relief effort. There are many families living in dispersed communities or in communities to which physical access is difficult. Community members are in the best position to render humanitarian aid to marooned victims. Communities are therefore important partners in risk reduction. They are the disaster front and must have the capacity to respond. To do this they must be involved in every stage of the disaster cycle – mitigation and preparedness, response and recovery.

Community leaders can educate. There are many practices that increase risks – disposal of garbage, removal of vegetation. The communities possess the type of information that allows micro planning. They can identify the location of heavy equipment, the elderly and the disabled. They can mobilize local teams to respond rapidly to crisis situations at each stage. They send out warnings, they organize teams to protect homes from flood waters using, for example sandbags, assist in rescue efforts. The preparation of shelters, the provision of emergency supplies are very important community activities.

Communities can lead the response phase. They can lead the rescue and are important in the recovery effort. Where the need is great, many community members provide temporary shelter. Neighbours form excellent support networks.

The Office of Disaster Preparedness in Jamaica has established disaster response mechanisms at three levels – National, Parish and community. The Parish organization mirrors the National, and is supported by a Parish Disaster Coordinator. At the community level there is a zonal programme in which communities with similar characteristics and within the sphere of influence of a growth centre are brought together to administer their disaster management needs.

Individual responses

The manner in which individuals respond to hazards depends on a number of factors. Many who live in hazardous areas adopt mitigation and preparedness strategies. In flood prone areas, houses are built on stilts; property is covered by insurance. The roofs of houses are constructed to withstand hurricane force winds. Decisions are made as to how important documents and household items are to be protected. Some make no preparation and bear the losses.

Before the event

Warnings

Before responding, individuals appraise the likelihood of the event and their actions depend on the level of confidence in the warning given. The false alarm rate for tsunamis in the Pacific region is 75 per cent. This reduces confidence in future warnings and if the warning is verified, the response may be too late to avoid casualties.

Perception of the hazard and its consequences

Their actions also depend on their perception of the hazard – whether it is likely to cause damage or mere discomfort and inconvenience which they are able to tolerate. Their perception is conditioned by previous experience.

Alternatives

Individuals weigh the possibility or the desirability of taking action and they are influenced by economic, social and cultural conditions. The issues confronting the people of New Orleans at the approach of Hurricane Katrina in 2005 are instructive. Many refused to heed the mandatory evacuation order because:

- There was confusion over the evacuation orders.
- They had religious faith.
- In the evacuation for Hurricane Ivan the previous year, many had fallen ill while stalled in vehicles.
- The hurricane occurred at the end of the month before they had received salaries and many could not afford to leave.
- Many had no transportation and could not afford transportation.
- Some felt that they had to protect their property in a high crime environment. There was no confidence in the ability of the police to protect property.
- The evacuation effort was too difficult for the sick and elderly.
- They were warned that alternative accommodation at the Louisiana Superdome would be uncomfortable.

Over 100,000 persons remained in the city and it is estimated that almost 2,000 persons died.

Personality

Some are risk takers and may regard the experience of a hazardous event as an adventure. Men are more likely to take risks than women and those with families are less likely then those without.

Activity 14.1

Describe three measures that communities can adopt to prepare themselves for the effects of disaster.

Feedback

Communities can prepare for the effects of disaster by public education, awareness and planning. Raised awareness of the hazard threat encourages individual and community response. While individuals may respond according to their perception of the threat as 'danger' or 'adventure', community leaders should stress the responsibility of individuals to avoid the type of environmental degradation that increases risk to each other in lessening the impact of any hazard on the community.

Community warning systems could be put in place.

Individuals could be encouraged to make preparations for the safety of the members of their own households and their vulnerable neighbours. Households with lifting machinery, sturdy means of transportation should be put on notice.

After the Event

Actions after the event are influenced by some of the same factors at work before the assault.

Perceptions about the event are important – Is it a rare event? What is the likelihood of another event in the person's lifetime? Is it likely to occur annually? The answers to these questions will determine expenditure on defences or decisions to relocate or to do nothing.

Alternatives. What options are available for relocation? Is it possible or affordable? Is modification of existing conditions more affordable than relocation? Is the investment in the hazardous location too high to be abandoned? Is bearing or sharing (insurance) the loss the only option?

Case Studies

The following section examines the response to flooding in Guyana in 2005; the earthquake in Costa Rica in 2009 and the eruption of the volcano in Montserrat in 1995. They each overwhelmed local resources and triggered regional and international responses.

Guyana floods, 2005

The extreme rainfall event in January 2005 and the flood impact on Guyana were described in Module 13 – Floods, Earthquakes and Volcanoes as Hazards. The most heavily populated and economically productive areas of coastal Guyana were covered in more than 2m of water. The government response was to immediately declare the three regions affected as disaster areas. The President and his Cabinet met with the Opposition leaders and the City Council to plan the response. Later meetings were held with the Guyana Red Cross and other non-governmental organizations. In the worse hit areas, the focus was on providing water, food and shelter for those affected. The government distributed thousands of hot meals, food hampers and bottles of water, 50 shelters were opened to house 4,200 persons.

After the flood waters receded, 350 health teams were sent to the area to prevent the spread of diseases. Other teams monitored the water levels in the large dams which had over topped during the flood. Sluice gates were opened to allow excess water from the Demarara River to ease out. Pumps were deployed at strategic points. Later donors assisted with more than US$ 800,000 in cash and US$ 1.6m in supplies to the Civil Defense Commission. Massive cleanup efforts were done to restore the mud covered buildings in the areas.

Recovery efforts are long term to mitigate the impact of the perennial flood hazard. Both structural and non structural methods are being used to mitigate further flooding. The drainage system is being strengthened. The Private Sector commission was asked to assist persons affected by the disaster. It is to be assumed that flood insurance would be part of any long term planning.

Costa Rica earthquake, 2009

Costa Rica is highly vulnerable to the three natural hazards mentioned here. It is in a very active tectonic area. It therefore needs to focus on preparedness and disaster risk reduction strategies. The National Emergency Commission (CNE) activated an Emergency Operations Centre to coordinate response measures after the earthquake of January

2009. The Costa Rican Red Cross and national fire fighters had search and rescue teams to take care of the injured and rescue stranded persons on damaged roads. The Costa Rica Social Service treated the injured at shelters and controlled sanitary conditions.

There are plans to create long term accommodation for some communities where practically all the buildings were destroyed. There will also have to be decisions on whether to rebuild and the use of seismic building designs to mitigate future events. The Costa Rican Volcanic and Seismic Observatory would need to strengthen prediction and communication systems to these remote areas.

Montserrat volcanic eruption, 1995

This British dependency depended on a response from the distant British government through their appointed Governor. Persons were evacuated from the south of the islands and offered safe passage to Britain and surrounding islands. In the long term the Governor is optimistic about the potential for adventure tourism and eventually building material and fertile soils.

Key Points

- Human settlement occurs in hazardous areas.
- Individual responses vary with economic, social and cultural conditions.
- Governments have the primary responsibility for managing disasters but individuals and communities must be encouraged to respond at each stage.

Conclusion

The growth of human populations continues to increase and so will the demand for living space. Most of the world is subject to one or more hazard risk. Therefore at all levels, individuals, communities and governments need to implement both structural and non structural strategies to reduce hazard impact. Improved prediction and technology may lessen the deaths from extreme natural events.

Activity 14.2

Describe three post-disaster activities that should be undertaken by government authorities as immediate responses to earthquakes.

Feedback

The first and immediate government response is to dispatch search and rescue emergency crews to try to save lives. Heavy equipment may be needed to remove collapsed buildings and dogs to sniff out those injured persons buried under the rubble. There would also have to be a needs assessment to ensure that appropriate supplies are delivered to those most in need. Food, water, shelter and sanitation facilities would all need to be put in place to avoid secondary effects such as famine and disease. Finally the government would need to restore law and order to the community of survivors. This would be recovery stage of hazard management both long and short term measures.

End Test

Governments can respond to the risk of hazards by adopting measures that avoid and that reduce risk. Describe three measures that avoid risk and two measures that reduce risk.

End Test Feedback

Three measures that avoid risk are land use regulations, public education and financial incentives. Governments can avoid hazard risks by restricting use of unsuitable or most hazardous areas. It may specify the safety limits in construction of roads and bridges.

Public education can raise awareness of the hazard risk so that individuals and communities are motivated to avoid the risk in their own decisions.

Financial incentives can also be used to avoid hazard risk. Tax exemptions for risk avoiding measures, infrastructure construction methods could encourage the private sector to be involved.

Suggested Further Reading

Bishop, V. (1998) *Hazards and Response*, London: Collins.

Guiness, P. and G. Nagle (2002) *Advanced Geography: Concepts and Cases*, London: Hodder and Stoughton.

Nagle, G. (1998) *Hazards*, London: Nelson.

Nagle, G. (1998) *Changing Settlements*, London: Nelson.

Nagle, G. (2000) *Advanced Geography*, Oxford: OUP.

Witherick, M. and K. Adams (2006) *Cities and Urbanisation*, London: Philip Allan.

Richardson, D. and St. John, P. (1997) *Methods of Presenting Fieldwork Data*, UK: The Geographical Association.

Strahler, A. and A. Strahler (2002) *Introducing Physical Geography* (Third Edition), New York: John Wiley and Sons.

Waugh, D. (2009) *Geography: An Integrated Approach* (Fourth Edition), Cheltenham: Nelson Thornes.